NEUROLOGY - LABORATORY AND CLINICAL RESEARCH DEVELOPMENTS

ALUMINUM NEUROTOXICITY

FROM SUBTLE MOLECULAR LESIONS TO NEUROLOGICAL DISEASES

NEUROLOGY - LABORATORY AND CLINICAL RESEARCH DEVELOPMENTS

Additional books in this series can be found on Nova's website
under the Series tab.

Additional e-books in this series can be found on Nova's website
under the eBooks tab.

NEUROLOGY - LABORATORY AND CLINICAL RESEARCH DEVELOPMENTS

ALUMINUM NEUROTOXICITY

FROM SUBTLE MOLECULAR LESIONS TO NEUROLOGICAL DISEASES

ANKICA JELENKOVIĆ
EDITOR

NOVA BIOMEDICAL
New York

Library of Congress Cataloging-in-Publication Data

Library of Congress Control Number: 2016931592

ISBN: 978-1-63484-739-1

Published by Nova Science Publishers, Inc. † New York

CONTENTS

PREFACE

Aluminum is a widespread metal in the earth. It is not an essential metal and does not appear to have biological functions in mammals. It is the scientific focus for many years as an environmental neurotoxic risk factor as it is a serious threat to human health. Exposure to aluminium in humans is primarily dietary, since aluminium compounds are added to many commercially-prepared products destined for ingestion. Also, aluminium is an ingredient of a number of drugs and vaccines. It is also used in the production of the dishes for preparing food. Moreover, some people are occupationally exposed to aluminum. Thus, exposure to environmental aluminium is difficult to avoid.

There is prolonged retention of aluminium that enters the brain, suggesting its potential for accumulation with repeated exposures. Its elimination is prolonged because of its very long elimination half-life and storage in more than one compartment. Neurotoxic effects of aluminium are complex and are manifested biochemically, structurally, and functionally. They are well documented in the human and in animal studies. Aluminium-induced disorders usually developed silently. Because of that, aluminium neurotoxicity is unlikely to be diagnosed in its early stage and remains clinically unapparent unless unmasked by experimental or natural processes.

Chapter 1 - In this chapter it is explained the neurotoxic effects of aluminium and mechanisms involved in mitochondrial oxidative stress generation under pathophysiology of aluminium neurotoxicity. It is well established that aluminium exerts toxic manifestations by exacerbating oxidative stress in the brain. Aluminium-induced neuronal oxidative stress and inhibition of the antioxidant defense system could be the possible mechanisms of its neurotoxicity. The oxidative stress has been implicated in pathogenesis of various neurodegenerative conditions including Alzheimer's disease and Parkinson's disease. Several studies have also suggested that mitochondrial oxidative damage is associated with the pathogenesis of these neurodegenerative disorders. Being involved in the production of reactive oxygen species, aluminium may impair mitochondrial bioenergetics and generate oxidative stress. In fact, aluminium has been shown to inhibit respiratory chain complexes activity and down regulate the mitochondrial antioxidant system. Further, evidence has also suggested that aluminium-induced oxidative damage to proteins during oxidative stress and proteins with specific sensitivity to oxidation can regulate cellular signaling events, including apoptosis.

Chapter 2 - Neurotransmission is essential to ensure activity-dependent plasticity of the central nervous system in physiological and pathophysiological conditions. Neuronal communication is a fine-tuned regulated process, which encompasses several strategies to enhance the paracrine and autocrine effects of signalling molecules. Gasotransmitters, the small free diffusible molecules, provide a way to shortcut from precursors to effectors pathways. Other neurotransmitters exploit sophisticated molecular apparatus, like exocytosis, the ancient cellular approach to reject poisoning molecules. Neuropeptide secretion is assured by calcium-dependent exocytosis, mediated by complexation of soluble *N*-ethylmaleimide-sensitive factor-attachment protein receptor (SNARE). Classical neurotransmitters also take advantage of other delivery routes for crossing membranous barriers, namely facilitated translocation by mediatophores and by the reversal of plasma membrane neurotransmitter transporters (PNTs). This chapter shows that the prevalence of aluminium at the synapse is a peculiar challenge for the regular synaptic function. Aside from dysregulation of intracellular signalling cascades by replacing regulatory cations in numberless molecular interactions, aluminium also interferes with supramolecular structures that are crucial for the release of neurotransmitters. PNTs are energetically coupled with the sodium/potassium-exchanging ATPase (Na/K-ATPase). PNTs are vulnerable to the aluminium-induced failure of neuronal Na/K-ATPase overactivation, which hinders PNTs –mediated neurotransmitter recycling at the synapse. Aluminium also jeopardizes the ultimate step of SNARE-mediated, calcium-dependent exocytosis, the dilation of the fusion pores through which vesicular pools of neurotransmitters are delivery to the synaptic cleft. Stimulus-evoked neurosecretion is diminished by aluminium. Hence, imbalance of neurotransmitter release should represent one of the cellular features of neuropathies exacerbated by increasing the brain burden of aluminium.

Chapter 3 - Neurotoxicological research is gaining strength at an increasing pace due to exploring and understanding of the multifold alterations in nervous system caused by environmental exposures to increasing pollutants and toxicants as well as pharmacological agents, leaving the issue of human health as a matter of great concern. Thus, mechanism of nervous regulation for learning and memory is importantly fragile to environmental exposures during early life stages that can potentially shift vulnerability at a later stage in life.

Although long-term exposure to environmental factors during pregnancy along with maternal stress can have serious negative effects on the learning abilities of the off spring, not much has been studied in human populations. However, such studies in altricial animals like rats and mice do provide relevant understanding from the perspective of human newborns. It is well understood that multiple exposure opportunities lie in these animals for early exposure of the fetus, which may include prenatal, postnatal and perinatal exposures, however; perinatal period is critically most important from the brain development point of view. Perinatal period in rodents is approximately equivalent to the first and second trimesters of human pregnancy, with the first 10 days or so of postnatal life in rats, approximating the third human trimester. Besides multiple exposure opportunities, the routes or modes of exposure is also an important factor that affects the developmental neurotoxicity and naturally oral administration appears to be relevantly and readily most common in human populations. Thus, the effect of perinatal oral exposure of environmental factors to pregnant mice needs to be assessed in the offspring at neurobehavioral, cognitive and biochemical levels (brain neurotransmitters and oxidative stress indices) to investigate the possible long term physiological mechanism associated with developmental neurotoxicity in the offspring at various postnatal developing stages.

Chapter 4 - The pathogenetic mechanisms of aluminum neurotoxicity have been the subject of various studies, with the ultimate goal aimed towards finding treatment options. In the authors' study, the involvement of nitric oxide (NO) in physiological and pathological processes in the brain is thoroughly discussed and it was experimentally examined. Under pathological conditions, this free radical may cause nitrosative stress and, consequently, provoke brain damage. Namely, rats were intrahippocampally treated with aluminum. Three doses of Nω-nitro-L-arginine methyl ester (L-NAME), the non-selective inhibitor of nitric oxide synthase, were applied before aluminum, while only the largest dose was given without aluminum pretreatment. Considering the index of lipid peroxidation and reduced glutathione in the four specific structures of the brain (forebrain cortex, striatum, basal forebrain and hippocampus), a wrong conclusion could have been made about the lack of involvement of NO in the pathogenesis of aluminum neurotoxicity, since the two lower doses of L-NAME didn't show any effects or their effects were very weak even though they reduced the effects of aluminum. However, apart from the fact that aluminum induced oxidative stress, it was discovered that NO was crucial for the development of oxidative stress and, also, that the reversion of aluminium induced disorders occurred at the "optimal" inhibition of its synthesis. The aforementioned was achieved only with the highest dose of L-NAME, which was without effects in rats that were not aluminum pretreated. In fact, it turned out that L-NAME showed protective effects in the neurotoxicity of aluminum, which were dose-dependent.

Chapter 5 - Neurological disorders which consociated with undue level of neuronal aluminum are revisited in this chapter in the light of aluminium toxicity. In the process of discussion, toxicodynamics of aluminum in reference to brain and spinal cord and possible implications of aluminium exposure faced by them in different pathophysiological conditions with their possible clinical correlations. Symptoms of both chronic and acute aluminum toxicities have been enlisted, as well as a list of neurological disorders where presence of aluminum is well established. A pictorial presentation of likely sequence of symptoms appearing and leading to increased morbidity in aluminum intoxication should facilitate understanding the events associated with aluminum-induced neurological insults. The chapter also emphasizes the major molecular targets of aluminum aggress especially in neurological disorders. Macroscopic, microscopic and ultrastructural changes are reviewed and linked with demonstrable biochemical and biopotential changes in human and animal studies. While dealing with neurological disorders associated with aluminum toxicity, the pathological conditions are tabulated as per the timescale of appearance of the neurological symptoms and identified as per the involvement of brain and spinal cord regions in observed neuropathology. In addition, neurological correlates between aluminum exposure and pathophysiology of Alzheimer's, Parkinson's and Huntington's disease, autism spectrum disorders, amyotrophic lateral sclerosis, gulf war syndrome and Fahr's diseases are discussed. Considering the controversial reports available regarding aluminum's role in neurological disorders, finally it is concluded that irrespective of aluminum's role in the neurological disorder, efforts should be made to minimize the exposure to aluminum, as its essentiality in our body is not proven yet.

Chapter 6 - Aluminium can induce inflammation, production of free radicals and oxidative modification of proteins and other biomolecules that lead to neurological disorders. The pathological features of aluminium neurotoxicity can be the same or can resemble the changes found in neurological diseases. As for example, aggregation of amyloidal plaques of beta-amyloid peptide extracellular or in the walls of cerebral blood vessels, and generation of neurofibrillary tangles in the neurons and glial cells, the characteristic histological lesions of Alzheimer's disease, are found in aluminium intoxication. Because of that, many current investigations are directed to the understanding the molecular mechanisms of aluminium neurotoxicity and to the associated neuropathological and other lesions. In this chapter it is overviewed the effects of Al on the central nervous system and the link between various neurodegenerative diseases. In particular, the conformational changes of amyloidogenic proteins and metal-metal interactions at the synapse. It is also discussed the fate of aluminium in the body.

Chapter 7 - Metal induced neurotoxicity has been reported to cause cognitive impairment in humans and animals. Metal dyshomeostasis has been implicated in the pathophysiology of various neurodegenerative diseases. Metal intoxication causes deleterious effects by activating redox sensitive signaling pathways involving mitogen activated protein (MAP) kinases and nuclear factor κ-B (NF- κ-B). This results in free radical generation causing oxidative stress. The intensification of neuroinflammation and the interference with cholinergic projection functions may represent the way by which aluminium contributes to pathological processes in cognitive dysfunction. There is a need to develop safe and effective therapy which can be protective against metal induced neurotoxicity. Antioxidants could be a promising alternate to chelation therapy as they are clinically safe and possess multiple mechanisms of actions. In fact, antioxidants (both enzymatic and non-enzymatic) have been reported to provide protection against deleterious metal mediated free radical attack. However, the neuroprotective effect of various antioxidants like curcumin, naringin, carvedilol remains unexplored against aluminium induced neurotoxicity and cognitive dysfunction. The present work has been designed to review the possible effects of antioxidants against metal induced cognitive dysfunction and oxidative stress.

Due to the similarity between neuropathological, biochemical and clinical findings in neurodegenerative and other brain diseases sufferers and aluminium overload, it is suggested that aluminium is probably involved in one or more aspects of the etiology/ pathogenesis/precipitation of them. Because of that and because of the large presence of aluminum in the environment, this book will be of interest to wide variety of scientists and professionals in different fields of science.

Preface was prepared by Ankica Jelenković, Editor,
and by authors of book chapters

In: Aluminum Neurotoxicity
Editor: Ankica Jelenković
ISBN: 978-1-63484-739-1

Chapter 1

MECHANISTIC ASPECTS OF ALUMINIUM NEUROTOXICITY: OXIDATIVE STRESS AND MITOCHONDRIAL DYSFUNCTIONS

***Vijay Kumar*[1] *and Kiran Dip Gill*[2,*]**
[1]Department of Biochemistry, Maharshi Dayanand University, Rohtak, India
[2]Department of Biochemistry, Postgraduate Institute of Medical Education and Research, Mohali, Punjab, India

ABSTRACT

Aluminum is present widely in the environment without showing any known biological function in human beings. The widespread use of aluminium containing products is increasing its bioavailability in our body. Although aluminium does not present with any noticeable toxicity to humans at low concentrations, but chronic exposure to this light toxic metal could play a role in increasing the vulnerability of the central nervous system to neuronal cell death. Its prooxidant effect in the brain becomes important since the nervous system is highly sensitive to oxidative damage resulting from free radicals and neurons have an elevated metabolic rate also. The generation of oxidative stress may be attributed to its toxic consequences in animals and humans. Oxidative stress is accompanied by generation of reactive oxygen species (ROS). It is widely accepted now that aluminium might exert its toxic effects through ROS generation. Mitochondria, being the major source of ROS generation may present an important target of aluminium neurotoxicity. Aluminium disturbs certain cell signaling cascades and induces neuronal apoptosis by its effect on the functioning of mitochondria leading to impaired mitochondrial bioenergetics and generation of oxidative damage in brain.

Keywords: aluminium, oxidative stress, neurotoxicity, mitochondria, neurodegeneration

* Corresponding author: Kiran Dip Gill. Dept. of Biochemistry, Postgraduate Institute of Medical, Education and Research, Chandigarh, India-160012. E-mail kdgill2002@yahoo.co.in, fax: 91-0172-2744401, 2745078; 4667 676.

INTRODUCTION

Aluminium is widely distributed in the environment comprising approximately 8% of the earth's crust. It is present mainly in soil, clay, minerals, rocks, water and food items. Industrial and occupational sources of exposure also contribute significantly to the atmospheric aluminium [1, 2]. Aluminium is used in preparation and storage of food in vessels, cans and foil, which may increase its content in foods [3]. The aluminium absorption is enhanced by the presence of citrate or low pH [4]. Hence, taking antacids along with fruit juice can be harmful. The increasing use of aluminium containing products is ensuring its burden in human body. Once absorbed, aluminium remains in the blood for several hours and subsequently eliminated either through the urinary tract or distributed in the body by the iron-binding protein transferrin [5] leading to its accumulation in various organs such as bone, lung, muscle, liver, and brain. The estimated average human consumption of aluminium from food and water varies between 2-30 mg [6] and only 10% of the ingested aluminium is absorbed in the body [7].

Aluminium is considered non-essential metal since it does not appear to have any biological function in animals and humans. However, its increased concentration causes both acute and chronic diseases in humans [8-10]. The high levels of aluminium has been associated with increased risk of a number of pathogenic disorders, such as microcytic anemia, osteomalacia [11, 12] and neurodegenerative disorders including dialysis encephalopathy, Alzheimer's disease (AD) and Parkinson's disease (PD) [8, 13]. Aluminium-induced neurotoxic manifestations have been shown in terms of neuropathological, neurobehavioral, neurophysical and neurochemical changes in rodents [14-18]. Aluminium uptake in the brain is much slower as compared to other organs but once gained access into the brain, it distributes into the various regions of brain [16]. The brain is considered to be most vulnerable to the toxic manifestations of aluminium owing to the non-dividing nature of the neurons, increased level of free radicals, low concentration of the endogenous glutathione and other antioxidants. The high proportion of polyunsaturated fatty acids makes this organ particularly susceptible to oxidative damage [19]. Aluminium is considered to have a strong prooxidant activity in spite of its non-redox status [20]. It causes oxidative damage by binding to pro-oxidant metals like iron, copper and modulates their ability to promote oxidative events. In fact, the mechanism postulated for aluminium-induced neurotoxicity majorly involves reactive oxygen species (ROS) mediated lipid peroxidation (LPO) and oxidative damage to proteins and DNA and decrease in intracellular glutathione [21-23].

Increasing evidence in recent years has suggested that mitochondrial dysfunction may play an important role in the pathogenesis of many neurodegenerative disorders [24]. Aluminium has been reported to have adverse effects on mitochondrial functions. It has been suggested that chronic exposure to aluminum causes the impairment of mitochondrial energy metabolism in different region of rat brain, which leads to progressive deterioration of mitochondrial functions and culminates into excessive free radical generation eventually resulting in DNA damage [21]. It can inhibit pyruvate utilization and lead to accumulation of citrate and oxoglutarate, decrease mitochondrial acetyl-CoA, and increase total mitochondrial calcium [25]. Aluminium has been shown to inhibit respiratory chain complexes activity and downregulation of mitochondrial antioxidant system. This chapter focuses on the disparate of aluminium on the brain with particular emphasis on mitochondrial oxidative stress and dysfunction. The chapter begins with neurotoxic effects of aluminium followed by discussion on mechanisms of mitochondrial oxidative stress generation under pathophysiology of aluminium neurotoxicity.

NEUROTOXICITY OF ALUMINIUM

The widespread presence of aluminium in the environment and its augmented use in the food processing, storage and pharmaceutical industry, has led to potential interest in its varied neurotoxic effects. Some of the neurotoxic symptoms of aluminium exposure include speech disturbances, dysparaxia, tremors, psychosis, personality changes or partial paralysis [26]. Brain becomes a soft target for aluminium toxicity as it freely crosses blood-brain barrier. Aluminium salts have been shown to impair cognitive functions in animals and humans. Acute and sub-acute exposure of aluminium to rats lead its accumulation in brain; maximum being in hippocampus [16]. High environmental levels of this element have been associated with pathological changes in central nervous system in the form of neurofibrillary tangles [27] and amyloid-β peptide and senile plaques [28]. Aluminium complexing can enhance the amyloid-β protein penetration of blood brain barrier [29]. Aluminium causes the accumulation of tau protein and amyloid-β protein in experimental animals and induces neural apoptosis *in vivo* as well as *in vitro* [30]. Crapper et al. [31] have also reported that aluminium gets distributed in the brain following ingestion in AD patients and lead to neurofibrillary degeneration. Tau is a microtubule-associated protein that requires phosphate for its function. The hyperphosphorylation of tau causes loss of its biological activity and proceeds to the formation of neurofibrillary tangles in brains of AD patients. Moreover, aluminium exposure may also induce neurotoxicity by causing destabilization and disruption of cytoskeletal framework, eventually leading to axonal degeneration since chronic aluminium exposure has been implicated in neurofilament phosphorylation and disruption of microtubules in Wistar rats [16]. Hence, aluminium may affect the peripheral nervous system by disrupting the neuroskeletal integrity and diminishing the nerve conduction velocity in the peripheral nerves to cause peripheral neuropathy. Studies have shown aluminium to interact with the cholinergic system leading to altered cholinergic projection functioning and intensifying its inflammation. The decrease in acetylcholinesterase activity with altered mental status and weakness can be associated with peripheral neuropathy, neuropsychiatric abnormalities, and extrapyramidal disorders. The interference with cholinergic projection functions may represent another way by which aluminium contributes to pathological processes in AD leading to learning and memory deficits [32, 33]. Aluminium is also known to produce neurotoxic symptoms in humans also. Neurological dysfunctions in patients of dialysis encephalopathy have been related to high aluminium concentrations in the dialysate and use of phosphate binding gels containing high amount of aluminium [34]. Neurotoxic effects of aluminium among foundry workers were linked to AD by Polizzie et al. [2]. The development of disorientation, memory disturbance, emotional incontinence, general convulsions and consciousness disturbance were reported in a case of aluminium encephalopathy with chronic renal failure following a 15 year dosage regimen of 3.0 g hydroxy-Al gel/day by Shirabe et al. [35]. Neuropathologically, the brain showed nerve cell atrophy and mild loss with stromal spongiosis, proliferation of astrocytes and microglia in the cerebral cortex, basal ganglia and thalamus. The proposed mechanisms of aluminium-induced neurotoxicity have been summarized in Figure 1.

Oxidative Stress in Aluminium Neurotoxicity

Oxidative stress is a condition associated with the generation of free radicals as a result of toxic insults. In normal conditions also, free radicals are generated inside the cell but the intracellular antioxidant enzyme readily scavenge them. An imbalance in generation and removal of free radicals produces oxidative stress. All aerobic cells suffer oxidative damage but the mammalian brain is considered to be highly sensitive [36] due to high O_2 consumption. In adult humans, the brain accounts for only a few percent of body weight, but consume about 20% of basal O_2 intake. This high O_2 uptake is to cope with the vast amounts of ATP needed to maintain neuronal intracellular ion homeostasis associated with propagation of action potentials and neurosecretion.

Oxidative stress is involved in development of neurodegenerative diseases, where damage to neurons can reflect both an increase in oxidative processes and a decrease in antioxidant defenses. Several evidence has shown the presence of oxidation products of proteins, lipids and DNA in postmortem tissue from AD and PD patients [37, 38] suggesting role of oxidative stress in these diseases. Although aluminium is not a transition metal but increased ROS generation has been reported during aluminium exposure [39, 40] and this increased ROS play major role in the onset of aluminium neurotoxicity [41, 42]. The ROS may subsequently attack almost all cell components including membrane lipids. Studies have also shown that aluminium compounds can enhance the LPO in liposomes and change membrane fluidity [43]. Aluminium addition to rat brain synaptosomes increased its accumulation and formation of thiobarbituric acid reactive substances suggesting that facilitation of aluminium accumulation during brain oxidative injury might contribute to aluminium neurotoxicity and neuronal cell degeneration [44]. Another study by Verstraeten et al. [45] showed that myelin is preferred target of aluminium mediated oxidative damage which may be attributed to high lipid content of myelin. Riu and Yongjian [46] depicted accumulation of oxidative damage in hippocampus and cerebral cortex of mice brain following aluminium chloride (10, 50 or 300 mg Al/kg b wt) administration for 100 days. The decrease in enzyme activities could be related to a reduced synthesis of the enzyme proteins and intracellular antioxidants as a result of aluminium neurotoxicity. Glutathione is one such intracellular antioxidant and can interact directly with ROS to detoxify them. Aluminium is also known to cause increase in LPO in the mice brain [33]. The elevation of LPO suggests participation of free radical-induced oxidative cell injury in mediating neurotoxicity of aluminium. LPO of biological membranes may result in the loss of membrane fluidity, changes in membrane potential and increase in membrane permeability.

Mitochondrial Oxidative Stress and Dysfunction in Aluminium Neurotoxicity

Mitochondria consume about 90% of the body's oxygen and are the richest endogenous source of ROS, since the redox centers of the mitochondrial electron transport chain constitute the major source of cellular superoxide production as a by-product of oxidative phosphorylation [47, 48]. Chronic ROS exposure can result in oxidative damage to mitochondrial and cellular proteins, lipids and DNA. Oxidative damage accumulates more in mitochondria than rest of the cells because electrons continually leak from the respiratory chain to form damaging ROS [49]. A growing body of evidence points to a central role of oxidative stress and mitochondrial dysfunction in neurodegenerative disease pathophysiology [48, 50, 51]. Mitochondrial DNA is susceptible to oxidative damage and increased oxidative damage to the inner membrane of the mitochondria leads to imbalances in the electron transport chain, resulting in increased superoxide and hydrogen peroxide production, which, in turn, further damages membrane proteins [52]. Also, acute ROS exposure can inactivate the iron sulphur centers of ETC complexes I, II, III and TCA cycle aconitase resulting in decreased mitochondrial energy production [53]. Mitochondrial dysfunction may cause wide range of signs and symptoms of structural and functional abnormalities in these organs [54]. Severe mitochondrial dysfunctions may lead to the opening of mitochondrial permeability transition pore and uncouple the oxidative phosphorylation; a phenomenon associated with impaired cellular ATP generation, increased DNA fragmentation and ultimately cell death [55]. It seems possible that aluminium by causing impairments in mitochondrial bioenergetics may be involved in the generation of ROS from mitochondria leading to induction of oxidative stress. Excessive generation of intracellular ROS produces neuronal damage by initiating direct damage to cellular macromolecules, perhaps by influencing gene expression and post-translational modifications of key elements of the apoptosis cascade [56]. Many studies have indicated that oxidative stress, release of calcium from intracellular stores, and perturbation of mitochondrial function may represent important steps in the mechanisms underlying neuronal cell death induced by aluminum [42, 57].

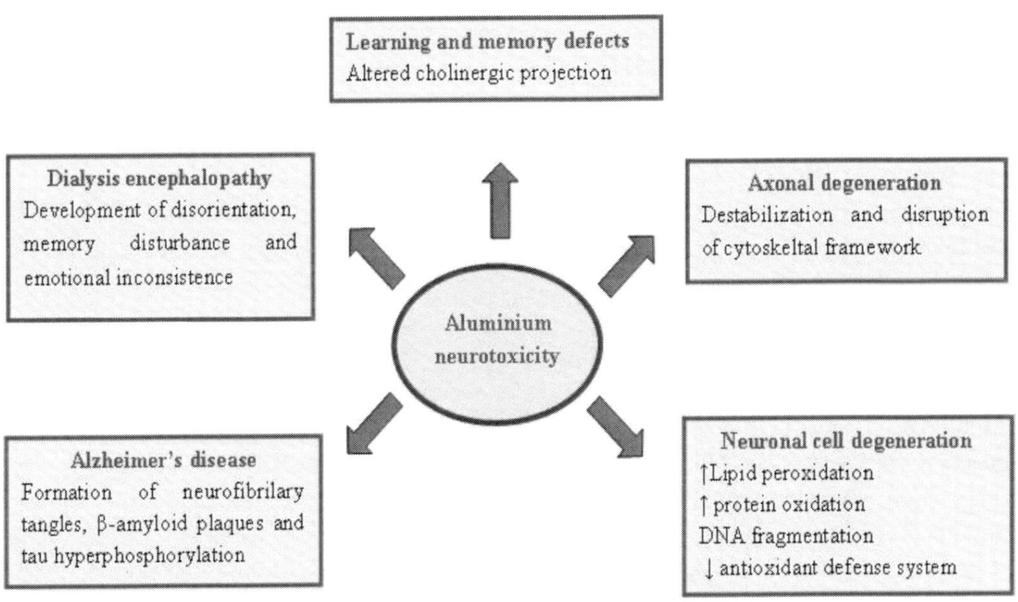

Figure 1. Mechanism of aluminium induced neurotoxicity.

Impaired Mitochondrial Energy Metabolism and Oxidative Phosphorylation

Free radicals generated through mitochondrial metabolism can act as causative factor for abnormal functions and cell death. Increasing evidence sustains the hypothesis that mitochondrial energy metabolism underlies the pathogenesis of neurodegenerative diseases. The postmortem brain samples of PD patients exhibited decreased complex I activity [58]. Similarly, impaired complex IV activity has been shown in AD patients [59]. Increased free radical mediated oxidative stress has been associated with the development of such disorders. A number of studies suggest that aluminium inhibits the mitochondrial respiratory system that may further elevate ROS production in neuronal cells. Aluminum has been reported to inhibit the complex I, II and IV activity in rat brain mitochondria [21, 60, 61] which suggests overall perturbation of electron transport chain leading to mitochondrial dysfunction. Mohan et al. [62] have reported decreased cytochrome *c* oxidase activity in brain mitochondria of New Zealand white rabbit exposed to aluminium (330 mg/kg/L) through drinking water. Aluminum exposure to human astrocytoma cells has been implicated with induction of mitochondrial dysfunction and decreased ATP production [63]. Chronic aluminium exposure also results in increased ROS generation, decreased MnSOD and aconitase activity in mitochondria from different region of rat brain [21, 22]. This decrease in mitochondrial antioxidant enzyme may be associated with increased mitochondrial protein oxidation in brain [22]. Sharma et al. [64] demonstrated decreased mRNA levels of the mitochondrial and nuclear encoded subunits. Besides, a decrease in mitochondrial DNA copy number, mitochondrial content and morphological changes in mitochondria in rat brain were also observed. There was down regulation of PGC-1α and its downstream targets. PGC-1α was suggested to be a potent target for aluminium neurotoxicity. Rats chronically exposed to aluminium also shows decline in ATP levels in brain mitochondria [21, 64]. Since ATP is the primary source of energy for the cell, its adequate concentration is essential for cell survival. The depletion of cellular ATP may cause deleterious effects on the ability of the cells to maintain ionic gradients through the function of ATP dependant translocases and could result in cell death [65]. De Marchi et al. [66] have also reported enhanced ROS production, decreased mitochondrial membrane potential and decreased activity of enzymes in mitochondria after aluminium toxicity. In fact, the free radical generation may target membrane lipids and proteins in proximity to respiratory chain and alter the fluidity and permeability of mitochondrial membrane.

Dysregulated Calcium (Ca^{2+}) Homeostasis

Altered intracellular free Ca^{2+} acts as a messenger and directly regulates growth, differentiation, and cell death [67]. Cellular metabolism and Ca^{2+} may be involved in activating the pathways leading to a loss of membrane integrity by change in lipid composition, cytoskeleton abnormalities and generation of free radicals. Mitochondrial dysfunction is directly associated with mitochondrial membrane permeability which regulates membrane potential [68]. Aluminium has been reported to decline the mitochondrial membrane potential in concentration dependent manner in rat neuronal cells [69]. Some studies of aluminium neurotoxicity have correlated mitochondrial dysfunction with intracellular Ca^{+2} levels. Impaired mitochondrial functions contribute to disturbed Ca^{2+} homeostasis by affecting buffering capacity of endoplasmic reticulum Ca^{2+} channels [70]. In one study, aluminium exposure have been reported to cause marked increase in the levels of intracellular Ca^{2+} level along with an increased LPO and a consequent decrease in fluidity of synaptic plasma membranes [71]. Chronic oral aluminium exposure significantly decreases total Ca^{2+}-ATPase activity in rat brain [72], disrupting Ca^{2+} transport from the cytoplasm. Drago et al. [73] have provided the evidence of contributing role of aluminium in the development of AD-related neuronal loss by impairing neuronal metabolism and Ca^{2+} homeostasis. Findings of a recent study demonstrated that chronic aluminum exposure resulted in spatial learning impairment and significantly increased intracellular Ca^{2+} level in the hippocampus of rats [74]. Extracellular aluminium inhibits Ca^{2+} influx into the neurons, which may impair acetyl-CoA transport from mitochondria to cytoplasm of nerve terminals causing suppression of acetylcholine synthesis and release [25, 75].

Mitochondria Mediated Apoptosis in Aluminum Neurotoxicity

Apoptotic cell death plays critical role in normal functioning and maintenance of tissue homeostasis which has been implicated with the process of cell death in brain enhanced by neurotoxic agents [76]. Cytotoxic stimuli mediated impaired mitochondrial integrity and functions represent a primary event in apoptotic cell death. Aluminium exposure induces oxidative stress and enhances mitochondria mediated apoptosis in neurons *in vivo* and *in vitro* [42].

ROS lead to increase in mitochondrial membrane permeability which may causes perturbation of mitochondrial functions [77]. The mitochondrial membrane permeability depends on mitochondrial permeability transition pore, a poly-protein complex which regulates the release of cytochrome *c* from the mitochondria to the cytosol [78]. In the cytosol cytochrome c binds with Apaf-1 and procaspase-9, and with the help of ATP they form an "apoptosome" [79] which leads to activation of caspase-9, an initiator of the cytochrome c dependent caspase cascade. During the mitochondria-mediated apoptotic process, Bcl-2 acts as anti-apoptotic factor and Bax as a pro-apoptotic factor. The balance between the pro- and anti-apoptotic proteins of the Bcl-2 family is important to determine cell survival or death. Murine neuroblastoma Neuro-2a cells treated with aluminium maltolate resulted in slight increase in caspase 3 activity along with down-regulation of Bcl-2 and an increased expression of Bax [57]. Intracisternal injection of aluminium-maltonate (50 μM) into rabbit brain was shown to induce cytochrome c release from mitochondria, decrease in mitochondrial Bcl-2 and caspase-3 activation suggesting that mitochondrial mediated apoptosis pathway and mitochondrial changes following aluminium-induced cytotoxic stimuli represent a primary event in apoptotic cell death [80]. Recently, Rizvi et al. [81] have reported increased mRNA and protein levels of caspase-3, caspase-9, Bax and cytochrome c but concomitantly increased levels of Bcl-2 in SH-SY5Y Neuroblastoma Cells treated with aluminium maltolate. Rats orally exposed to aluminium chloride (100 mg/kg b. wt.) for 8 weeks have shown significant increase in the protein expression of cytochrome c, Bax, Apaf-1 and caspases in brain [82]. The various mechanisms of aluminium-induced mitochondrial oxidative stress and dysfunction have been shown in Figure 2.

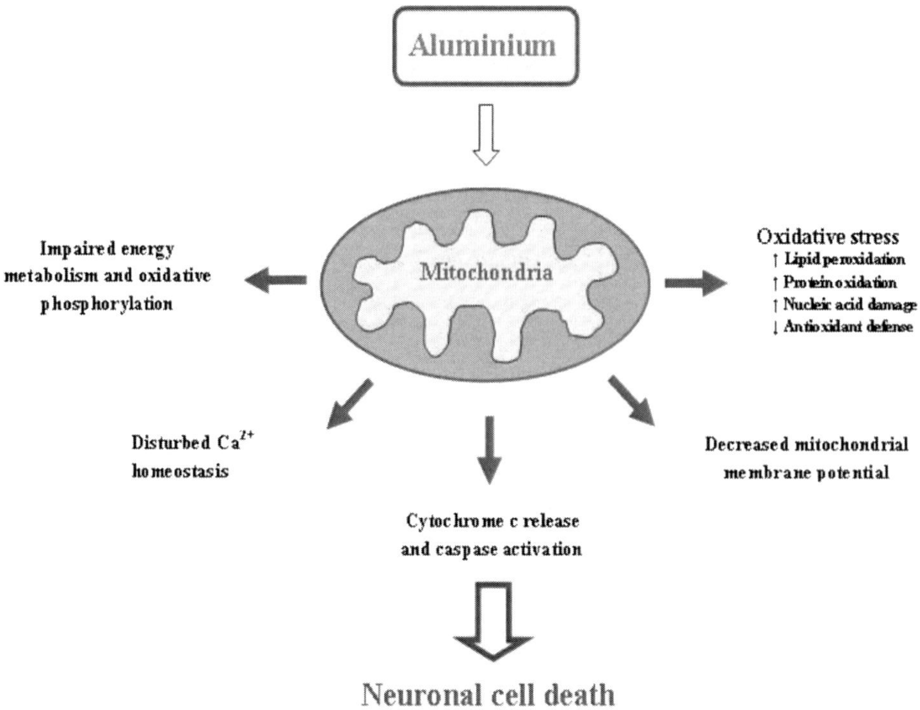

Figure 2. Mechanism of aluminium induced mitochondrial oxidative stress and dysfunction.

Conclusion

Aluminium is widely accepted as a neurotoxicant now and diverse mechanisms of its neurotoxicity have been presented so far. It is well established fact now that aluminium exerts toxic manifestations by exacerbating oxidative stress in brain. Moreover, mitochondria associated dysfunctions may also be regarded as causative factor for aluminium mediating toxicity. It has been proposed that it is the functional impairment of mitochondria that may be responsible for the generation of ROS which may be implicated in the oxidative damage to proteins and DNA and loss of its functions. This in turn may explain the possible involvement of aluminium in the pathogenesis of neurodegenerative diseases such as AD in which the presence of aggregated forms of some of the proteins is well established fact now.

References

[1] Buchta, M., Kiesswetter, E., Otto, A., Schaller, K.H., Seeber, A., Hilla, W. et al. 2003. Longitudinal study examining the neurotoxicity of occupational exposure to aluminium containing welding fumes. *Int. Arch. Occup. Environ. Health* 76, 539-558.
[2] Polizzi, S., Pira, E., Ferrara, M., Bugiani, M., Papaleo, A., Albera, R. et al. 2002. Neurotoxic effects of aluminium among foundry workers and Alzheimer's disease. *Neurotoxicology* 6, 761-774.

[3] Ribes, D., Colomina, M.T., Vicens, P., Domingo, J.L. 2008. Effects of oral aluminum exposure on behavior and neurogenesis in a transgenic mouse model of Alzheimer's disease. *Exp. Neurol.* 214, 293-300.

[4] Fulton, B., Jaw, S., Jeffery, E.H. 1989. Bioavailability of aluminum from drinking water. *Fundam. Appl. Toxicol.* 12, 144-150.

[5] Harris, W.R., Berthon, G., Day, J.P., Exley, C., Flaten, T.P., Forbes, W.F., et al. 1996. Speciation of aluminum in biological systems. *J. Toxicol. Environ. Health* 48, 543-568.

[6] Greger, J.L., Goetz, W., Sullivan, D. 1985. Aluminum levels in foods cooked and stored in aluminum pans, trays and foil. *J. Food Protect.* 48, 772-777.

[7] Gorsky, J.E., Dietz, A.A., Spencer, H., Osis, D. 1979. Metabolic balance of aluminium studied in six men. *Clin. Chem.* 25, 1739-1743.

[8] Yokel, R.A. 2000. The toxicology of aluminum in the brain: a review. *Neurotoxicology* 21, 813-828.

[9] Exley, C. 2001. Aluminium, Alzheimer's disease. *J. Alzheimers Dis.* 3, 551-552.

[10] Exley, C. 2005. The aluminium-amyloid cascade hypothesis and Alzheimer's disease. *Subcell. Biochem.* 38, 225-234. Doi: 10.1007/0-387-23226-5_11.

[11] Suwalsky, M., Norris, B., Villena, F., Cuevas, F., Sotomayor, P., Zatta, P. 2004. Aluminum fluoride affects the structure and functions of cell membranes. *Food Chem. Toxicol.* 42, 925-933.

[12] Domingo, J.L., 2006. Aluminum and other metals in Alzheimer's disease: a review of potential therapy with chelating agents. *J. Alzheimer's Dis.* 10,331-341.

[13] Becaria, A., Campbell, A., Bondy, S.C. 2002. Aluminum as a toxicant. *Toxicol. Ind. Health* 18, 309-320.

[14] Miu, A.C., Andreescu, C.E., Vasiu, R., Olteanu, A.I. 2003. A behavioral and histological study of the effects of long-term exposure of adult rats to aluminum. *Int. J Neurosci.* 113, 1197-1211.

[15] Colomina, M.T., Roig, J.L, Sanchez, D.J., Domingo, J.L. 2002. Influence of age on aluminum-induced neurobehavioral effects and morphological changes in rat brain. *Neurotoxicology* 23, 775-781.

[16] Kaur, A., Joshi, K., Minz, R.W., Gill, K.D. 2006. Neurofilament phosphorylation and disruption: a possible mechanism of chronic aluminium toxicity in Wistar rats. *Toxicology* 219, 1-10.

[17] Walton, J.R. 2012. Aluminum disruption of calcium homeostasis and signal transduction resembles change that occurs in aging and Alzheimer's disease. *J. Alzheimers Dis.* 29, 255-273.

[18] Walton, J.R. 2013. Aluminum involvement in the progression of Alzheimer's disease. *J. Alzheimers Dis.* 35, 7-43.

[19] Abdel Moneim, A.E. 2012. Flaxseed oil as a neuroprotective agent on lead acetate-induced monoamineric alterations and neurotoxicity in rats. *Biol. Trace Elem. Res.* 148, 363-370.

[20] Zatta, P., Kiss, T., Suwalsky, M., Berthon, G. 2002. Aluminium (III) as a promoter of cellular oxidation. *Coord. Chem. Rev.* 228, 271-284.

[21] Kumar, V., Bal, A., Gill, K.D. 2008. Impairment of mitochondrial energy metabolism in different regions of rat brain following chronic exposure to aluminium. *Brain Res.* 232, 94-103.

[22] Kumar, V., Bal, A., Gill, K.D. 2009. Aluminium-induced oxidative DNA damage recognition and cell-cycle disruption in different regions of rat brain. *Toxicology* 264, 137-144.

[23] Gonzalez, M.A., Bernal, C.A., Mahieu, S., Carrillo, M.C. 2009. The interactions between the chronic exposure to Aluminum and liver regeneration on bile flow and organic anion transport in rats. *Biol. Trace Elem. Res.* 127, 164-176.

[24] Leuner, K., Hauptmann, S., Abdel-Kader, R., Scherping, I., Keil, U., Strosznajder, J.B. 2007. Mitochondrial dysfunction: the first domino in brain aging and Alzheimer's disease? *Antioxid. Redox. Signal.* 9, 1659-1675.

[25] Szutowicz, A., Bielarczyk, H., Kisielevski, Y., Jankowska, A., Madziar, B., Tomaszewicz, M. 1998. Effects of aluminum and calcium on acetyl-CoA metabolism in rat brain mitochondria. *J. Neurochem.* 71, 2447-2453.

[26] Nayak, P. 2002. Aluminum: impacts and disease. *Environ. Res.* 89, 101-115.

[27] Perl, D.P., Gajdusek, D.C., Garruto, R.M. 1982. Intraneuronal aluminium accumulation in any itrophic lateral sclerosis and Parkinsonism: dementia of Gaum. *Science* 217, 1053-1055.

[28] Abramov, A.Y., Canevari, C., Duchen, M.R. 2004. Beta-amyloid peptides induce mitochondrial dysfunction and oxidative stress in astrocytes and death of neurons through activation of NADPH oxidase. *J. Neurosci.* 24, 565-575.

[29] Banks, W.A., Nieho, V., Drago, M.L., Zatta, D.P. 2006. Aluminium complexing enhances amyloid beta protein penetration of blood–brain barrier. *Brain Res.* 1116, 215-221.

[30] Kawahara, M. 2005. Effect of aluminium on the nervous system and its possible link with neurodegenerative diseases. *J. Alzheimers Dis.* 8, 171-182.

[31] Crapper McLachlan, D.R., DeBoni, U. 1980. Aluminium in human brain disease-an overview. *Neurotoxicology* 1, 3-16.

[32] Platt, B., Fiddler, G., Riedel, G., Henderson, Z. 2001. Aluminium toxicity in the rat brain: histochemical and immunocytochemical evidence. *Brain Res. Bull.* 55, 257-267.

[33] Kaizer, R.R., Correa M.C., Spanevello R.M., Morsch, V.M., Mazzanti C.M., Goncalves J.F. et al. 2005. Acetylcholinesterase activation and enhanced lipid peroxidation after long-term exposure to low levels of aluminum on different mouse brain regions. *J. Inorg. Biochem.* 99, 1865-70.

[34] Jack, R., Rabin, P.L., McKinney, T.D. 1984. Dialysis encephalopathy. A review. *Int. J. Psychiatr. Med.* 13, 309-326.

[35] Shirabe, T., Irie, K., Uchida, M. 2002. Autopsy case of aluminium encephalopathy. *Neuropathology*. 22, 206-210.

[36] Halliwell, B. 2006. Oxidative stress and neurodegeneration: where are we now? *J. Neurochem.* 97, 1634-1658.

[37] Giasson, B.I., Ischiropoulos, H., Lee, V.M., Trojanowski, J.Q. 2002. The relationship between oxidative/nitrative stress and pathological inclusions in Alzheimer's and Parkinson's diseases. *Free Radic. Biol. Med.* 32, 1264-1275.

[38] Jenner, P., 2003. Oxidative stress in Parkinson's disease. *Ann. Neurol.* 53, S26-S36 (discussion S36-S38).

[39] Strong, M.J., Garruto, R.M., Joshi, J.G., Mundy, W.R., Shafer, T.J. 1996. Can the mechanisms of aluminium neurotoxicity be integrated into a unified scheme? *J. Toxicol. Environ. Health* 48, 599-613.

[40] Prakash, N.T., Rao, K.S. 1995. Modulations in antioxidant enzymes in different tissues of marine bivalve Perna viridis during heavy metal exposure. *Mol. Cell Biochem.* 146, 107-113.
[41] Ohyashiki, T., Suzuki, S., Satoh, E., Uemori, Y. 1998. A marked stimulation of Fe2+-initiated lipid peroxidation in phospholipid liposomes by a lipophilic aluminum complex, aluminum acetylacetonate. *Biochim. Biophys. Acta.* 1389, 141-149.
[42] Savory, J., Herman, M.M., Ghribi, O. 2003. Intracellular mechanisms underlying aluminium- induced apoptosis in rabbit brain. *J. Inorg. Biochem.* 97, 151-154.
[43] Kaneko, N., Sugioka, T., Sakurai, H. 2007. Aluminium compounds enhance lipid peroxidation in liposomes: insight into cellular damage caused by oxidative stress. *J. Inorg. Biochem.* 101, 967-975.
[44] Amador, F.C., Santos, M.S., Oliveira, C.R. 1999. Lipid peroxidation facilitates aluminum accumulation in rat brain synaptosomes. *J. Toxicol. Environ. Health A* 58, 427-435.
[45] Verstraeten, S.V., Golub, M.S., Keen, C.L., Oteiza, P.I. 1997. Myelin is a preferential target of aluminum-mediated oxidative damage. *Arch. Biochem. Biophys.* 344, 289-294.
[46] Rui D, Yongjian Y. 2010. Aluminum chloride induced oxidative damage on cells derived from hippocampus and cortex of ICR mice. *Brain Res.* 1324:96-102.
[47] Cadenas, E., Davies, K.J.A. 2000. Mitochondrial free radical generation, oxidative stress, and aging. *Free Rad. Biol. Med.* 29, 222-230.
[48] Wallace, D.C. 2005. A mitochondrial paradigm of metabolic and degenerative diseases, aging, and cancer: a dawn for evolutionary medicine. *Annu. Rev. Genet.* 39, 359-407.
[49] Smith, R.A.J., Porteous, C.M., Murphy, C.M. 1999. Selective targeting of an antioxidant to mitochondria. *Eur. J. Biochem.* 263, 709-716.
[50] Andersen, J.K. 2004. Oxidative stress in neurodegeneration: cause or consequence? *Nat. Med.* 10, S18-S25.
[51] Beal, M.F. 2005. Mitochondria take center stage in aging and neurodegeneration. *Ann. Neurol.* 58, 495-505.
[52] Yates, F.M., Van Houten, B. 1997. Mitochondrial DNA damage is more extensive and persists longer than nuclear DNA damage in human cells following oxidative stress. *Proc. Natl. Acad. Sci. US* 94, 514-519.
[53] Wallace, D.C., Brown, M.D., Lott, M.T. 1999. Mitochondrial DNA variation in human evolution and disease. *Gene* 238, 211-230.
[54] Mehndiratta, M.M., Aggarwal, P., Singhal, R.K., Munjal, Y.P. 2000. Mitochondrial cytopathies. *J. Assoc. Phys. India* 48, 417-420.
[55] Le Masters, J.J., Nieminen, A.L., Qian, T., Trost, L.C., Elmore, S.P., Nishimura, Y. et al. 1998. The mitochondrial permeability transition in cell death: a common mechanism in necrosis, apoptosis and autophagy. *Biochim. Biophys. Acta.* 1366, 177-196.
[56] Sakaguchi, N., Inoue, M., Ogihara, Y. 1998. Reactive oxygen species and intracellular $Ca2^{+}$, common signals for apoptosis induced by Gallic acid. *Biochem. Pharmacol.* 5, 1973-1981.
[57] Johnson, V.J., Kim, S.H., Sharma, R.P. 2005. Aluminum-maltolate induces apoptosis and necrosis in neuro-2a cells: potential role for p53 signaling. *Toxicol. Sci.* 83, 329-239.
[58] Schapira, A.H.V. 1998. Mitochondrial dysfunction in neurodegenerative disorders. *Biochim. Biophys. Acta.* 1366, 225-233.

[59] Chagnon, P., Betard, C., Robitaille, Y., Cholette, A., Gauvreau, D. 1995. Distribution of brain cytochrome oxidase activity in various neurodegenerative diseases. *Neuro. Report* 6, 711-715.

[60] Sood, P.K., Nahar, U., Nehru, B. 2011. Curcumin attenuates aluminum-induced oxidative stress and mitochondrial dysfunction in rat brain. *Neurotox. Res.* 20, 351-361.

[61] Prakash A., Shur B., Kumar A. 2013. Naringin protects memory impairment and mitochondrial oxidative damage against aluminum-induced neurotoxicity in rats. *Int. J. Neurosci.* 123(9), 636-645.

[62] Mohan, N., Alleyne, T., Adogwa, A. 2009. The Effects of Ingested Aluminium on Brain Cytochrome Oxidase Activity. *West Indian Med. J.* 58, 422.

[63] Lemire, J., Mailloux, R., Puiseux-Dao, S., Appanna, V.D. 2009. Aluminum-induced defective mitochondrial metabolism perturbs cytoskeletal dynamics in human astrocytoma cells. *J. Neurosci. Res.* 87, 1474-1483.

[64] Sharma, D.R., Sunkaria, A., Wani, W.Y., Sharma, R.K., Kandimalla, R.J., Bal. A., et al. 2013. Aluminium induced oxidative stress results in decreased mitochondrial biogenesis via modu-lation of PGC-1a expression. *Toxicol. Appl. Pharmacol.* Doi: 10.1016/j.taap.2013.09.012 pii:S0041-008X(13)00412-2.

[65] Kraus-Friedman, N., Bibes, J., Muren, H., Carafoli, E. 1982. Calcium uptake in isolated hepatic plasma membrane vesicles. *Eur. J. Biochem.* 129, 7-12.

[66] De Marchi, U., Mancon, M., Battaglia, V., Ceccon, S., Cardellini, P., Toninello, A. 2004. Influence of reactive oxygen species production by monoamine oxidase activity on aluminum-induced mitochondrial permeability transition. *Cell Mol. Life Sci.* 61, 2664-2671.

[67] Orrenius, S., Zhivotovsky, B., Nicotera, P. 2003. Regulation of cell death: the calcium-apoptosis link. *Nat. Rev. Mol. Cell Biol.* 4, 552-565.

[68] Koya, R.C., Fujita. H., Shimizu, S., Ohtsu, M., Takimoto, M., Tsujimoto, Y. et al. 2000. Gelsolin inhibits apoptosis by blocking mitochondrial membrane potential loss and cytochrome c release. *J. Biol. Chem.* 19, 275, 15343-15349.

[69] Niu, P.Y., Niu, Q., Zhang, Q.L., Wang, L.P., He, S.E., Wu, T.C. et al. 2005. Aluminum impairs rat neural cell mitochondria in vitro. *Int. J. Immunopathol. Pharmacol.* 18, 683-689.

[70] Gibson, G.E., Hirsch, J.A., Cirio, R.T., Jordan, B.D., Fonzetti, P., Elder, J. 2013. Abnormal thiamine-dependent processes in Alzheimer's Disease. Lessons from diabetes. *Mol. Cell Neurosci.* 55, 17-25.

[71] Julka, D., Gill, K.D. 1996. Effect of aluminum on regional brain antioxidant defense status in Wistar rats. *Res. Exp. Med.* 196, 187-194.

[72] Julka, D., Gill, K.D. 1996. Altered calcium homeostasis: a possible mechanisms of aluminium-induced neurotoxicity. *Biochim. Biophys. Acta.* 1315: 47-54.

[73] Drago, D., Cavaliere, A., Mascetra, N., Ciavardelli, D., di Ilio, C., Zatta, P. et al. 2008. Aluminum modulates effects of beta amyloid(1-42) on neuronal calcium homeostasis and mitochondria functioning and is altered in a triple transgenic mouse model of Alzheimer's disease. *Rejuvenation Res.* 11, 861-871.

[74] Wang, B., Zhao, J., Yu, M., Meng, X., Cui, X., Zhao, Y. et al. 2014. Disturbance of intracellular calcium homeostasis and CaMKII/CREB signaling is associated with learning and memory impairments induced by chronic aluminum exposure. *Neurotox. Res.* 26, 52-63.

[75] Bielarczyk, H., Tomaszewicz, M., Szutowicz, A. 1998. Effect of aluminium on acetyl-CoA and acetylcholine metabolism in nerve terminals. *J. Neurochem.* 70, 1175-1181.

[76] Mattson, M.P. 2000. Apoptosis in neurodegenerative disorders. *Nat. Rev. Mol. Cell Biol.* 1, 120-129.

[77] Huang, X., Wullschleger, S., Shpiro, N., McGuire, V.A., Sakamoto, K., Woods, Y.L. et al. 2008. Important role of the LKB1-AMPK pathway in suppressing tumorigenesis in PTEN-deficient mice. *Biochem. J.* 412, 211-221.

[78] Liang, H., Ran, Q., Jang, Y.C., Holstein, D., Lechleiter, J., McDonald-Marsh, T. et al. 2009. Glutathione peroxidase 4 differentially regulates the release of apoptogenic proteins from mitochondria. *Free Radic. Biol. Med.* 47, 312-320.

[79] Love, S. 2003. Apoptosis and brain ischaemia. *Prog Neuropsychopharmacol. Biol. Psychiatry*. 27, 267-282.

[80] Ghribi, O., Herman, M.M., Forbes, M.S., DeWitt, D.A., Savory, J. 2001. GDNF protects against aluminum-induced apoptosis in rabbits by upregulating Bcl-2 and Bcl-XL and inhibiting mitochondrial Bax translocation. *Neurobiol. Dis.* 8, 764-773.

[81] Rizvi, S.H.M., Parveen, A., Verma, A.K., Ahmad, I., Arshad, M., Mahdi, A.A. 2014. Aluminium induced endoplasmic reticulum stress mediated cell death in SH-SY5Y neuroblastoma cell line is independent of p53. *PLoS One* 9, e98409.

[82] Singla, N., Dhawan, D.K. 2000. Zinc down regulates Apaf-1-dependent Bax/Bcl-2 mediated caspases activation during aluminium induced neurotoxicity. *Biometals* 28, 61-73.

In: Aluminum Neurotoxicity
Editor: Ankica Jelenković

ISBN: 978-1-63484-739-1

Chapter 2

ALUMINIUM AND IMBALANCE IN NEUROTRANSMITTER RELEASE

Paula P. Gonçalves[1,*], Robert Zorec[2,3] and Virgília S. Silva[1]

[1]CESAM and Departamento de Biologia, Universidade de Aveiro, Aveiro, Portugal

[2]Celica Biomedical, Ljubljana, Slovenia

[3]Laboratory of Neuroendocrinology-Molecular Cell Physiology, Faculty of Medicine, Institute of Pathophysiology, University of Ljubljana, Ljubljana, Slovenia

ABSTRACT

Aluminium, the third most abundant element in the earth's crust, is recognised as a neurotoxic agent. The properties of aluminium compounds and its alloys triggered their massive industrial applications. Aluminium bioavailability is increasing due to anthropological activities that include formation of aqueous soluble aluminium compounds. A body of evidence has established that the adverse effects of aluminium promote neurodegeneration in humans. Neurodegenerative diseases elevate morbidity and mortality, particularly in the elderly. There are many mechanisms of aluminium neurotoxicity, but it appears that they critically interfere with the main steps of neurotransmission at central nervous system synapses. To what extent aluminium induced effects on neurotransmission contribute to neuropathy, is the aim to be addressed in this chapter.

Keywords: aluminium-related neuropathies, aluminium neurotoxicity, neurotransmission

* Corresponding author: Paula P. Gonçalves. Departamento de Biologia, Universidade de Aveiro, 3810-193 Aveiro, Portugal. E-mail: pgoncalves@ua.pt.

Abbreviations

CBS	cystathionine-beta-synthase
CSE	cystathionine-gamma-lyase
GABA	γ-aminobutyric acid
HO	heme oxygenase
Na/K-ATPase	sodium/potassium-exchanging ATPase
NOS	nitric oxide synthase
PNTs	plasma membrane neurotransmitter transporters
SNARE	soluble *N*-ethylmaleimide-sensitive factorattachment protein receptor

Introduction

Existing naturally in complex compounds such as silicates and oxides, pure aluminium was isolated in the beginnings of the XIX century and the potential neurotoxic action of aluminium salts was even noted earlier (Altmann, 2001). The first experimentally demonstrated deleterious action of aluminium on the brain was neurodegeneration. Fifty years after the initial clinical report of aluminium poisoning with overt neurological symptoms by John Spofforth in *Lancet*, Allen C. Alfrey and colleagues unequivocally linked aluminium with the onset of dialysis encephalopathy (Alfrey et al., 1976). Consequently, this human misfortune has been minimized not only by controlling the quality of water used in haemodialysis, but also by circumventing the administration of aluminium containing medicaments to the patients and even detoxification with chelating agents (Percy et al., 2011). Now, we are approaching the end of the half-century during which aluminium has been implicated as a possible contributing factor in several neurological diseases that cause high morbidity and mortality. Since the mid-twentieth century, the controversial role of aluminium in the aetiology of neurodegenerative diseases, like Alzheimer's disease, affecting nearly 44 million people worldwide, is debated.

Clinical and epidemiological reports, supported by laboratory studies, of the capability of aluminium to root neuropathological conditions, bloomed recently. Evidence emphasises that human brains overloaded with aluminium show alterations that contribute to severe age-related neuropathies (Kawahara and Kato-Negishi, 2011; Exley, 2014; Kumar and Gill, 2014; Shaw et al., 2014). However, aluminium is still being described as a multitasking Janus ion, which interferes with different processes under pathophysiological conditions. Current hypotheses suggest aluminium-induced impairment of particular chemically coded neuronal populations. Nevertheless, its deleterious effects seem to be originating in the imbalance of a number of cellular processes. Aluminium might interfere with the main steps of neurotransmission (synthesis, storage, release, postsynaptic reception and inactivation of neurotransmitters). At the presynaptic nerve terminals, aluminium could modify neurotransmission in several ways: (i) by chelating neurotransmitter molecules (Deloncle and Guillard, 1990; De Stasio et al., 1994; Zhang et al., 2001); (ii) by modifying the activity of enzymes responsible for neurotransmitter synthesis and cleavage (Hofstetter et al., 1987; Nayak and Chatterjee, 2001); (iii) by compromising neurotransmitter release and recycling back to nerve terminals (Lai et al., 1981, 1982; Wong et al., 1981; Provan and Yokel, 1992; Struys-Ponsar et al., 2000; Nday et al., 2010); and (iv) by interfering with cell signalling through pre- and postsynaptic receptors (Julka et al., 1995; Trombley, 1998; Hermenegildo et al., 1999; Kim et al., 2007; Song et al., 2014).

Deterioration of cognitive abilities, such as loss of memory and behavioural learning, is a common feature of neurodegenerative diseases and aluminium toxicity (Kawahara and Kato-Negishi, 2011; Szutowicz, 2001; Kumar and Gill, 2014; Bondy, 2014; Shaw et al., 2014; Walton, 2014), for which dysregulation of synaptic plasticity is of foremost importance (Selkoe, 2002). Hippocampal long-term potentiation requires enhancement of synaptic efficiency and is closely associated with deficits in behavioural learning and memory. Thereby, it is not surprising that the effects of aluminium on long-term potentiation in hippocampus known for decades (Platt et al., 1995) warranted attention recently (e.g., Song et al., 2014; Zhang et al., 2014).

Aluminium is undoubtedly recognised as a neurotoxic agent. Recent works addressing the potential benefit of neuropharmacological products in prevention and attenuation of symptomatology associated with neurodegenerative conditions include the administration of aluminium compounds to experimental animals, given that aluminium induces neurodegeneration, probably as a result of direct toxic effects.

Excellent reviews have been entirely dedicated to aluminium related pathologies (Kawahara and Kato-Negishi, 2011; Bondy, 2014; Exley, 2014; Kumar and Gill, 2014; Pineau et al., 2014; Shaw et al., 2014; Walton, 2014; Willhite et al., 2014; Zhu et al., 2014), and the importance of neurotransmission impairment in aluminium neurotoxicity has been highlighted in the last years (Gonçalves and Silva, 2007; Verstraeten et al., 2008; Blaylock and Strunecká, 2009). There is a neurotransmission step scarcely considered so far; it assures the release of neurotransmitters to the synaptic cleft, which is the major focus of this chapter. Based on our own experience and accordingly with other authors, at least two complex molecular structures appear to escape from the dual mode of aluminium action, the electrogenic sodium/potassium-exchanging ATPase (Na/K-ATPase, EC 3.6.3.9), which moves two cation species in opposite directions across the plasma membrane, and the fusion pore that forms during regulated exocytosis. They will be herein considered as putative molecular targets in aluminium-related neuropathies through their intimate involvement in neurotransmitter release.

Aluminium-Related Neuropathies

In several occasions a direct association between metal exposure and anomalous neurological indicators was observed (Alfrey et al., 1976; Gherardi and Authier, 2012). Memory, concentration and speech deficits, impaired psychomotor control, reduced seizure tolerance and altered behaviour are manifestations of aluminium neurotoxicity, which have been linked to Alzheimer's and Parkinson's disease, amyotrophic lateral sclerosis, dementia, multiple sclerosis and other neurological impairments (Exley, 2014; Shaw et al., 2014; Walton, 2014). The most recognized health effects in humans from exposure to aluminium are summarized in Table I.

Table I. Representative health effects in humans from exposure to aluminium

Exposure pathway	Population/Sample	Aluminium source and/or compound	Dose and/or duration	Aluminium effects	Reference
Oral	Adults (elderly subjects)	Various dietary	Chronic	Impaired visuomotor coordination, poor long-term memory, and increased sensitivity to flicker (correlated with high serum aluminium level levels).	Bowdler et al., 1979.
Oral	Infant	Antacids containing aluminium	Chronic	Craniosynostosis (premature ossification of the skull and obliteration of the sutures).	Pivnick et al., 1995.
Oral	Adults (30 women and 25 men, aged 15-70 years old)	Aluminium sulphate (present as flocculant in potable water supplies, accidentally released in high amounts)	3 years after the accidental contamination of drinking water in Camelford	Symptoms of loss of concentration and short term memory. Psychomotor performance was poorer than predicted and the difference between flash-pattern visual evoked potentials was greater than normal. Long term cerebral impairment.	Altmann et al., 1999.
Oral	Adults (3,777 subjects aged 65 years old and older)	Drinking water supplies	Aluminium levels ranged from 0.001 to 0.459 mg/litre/8 years	Dementia and Alzheimer's disease.	Rondeau et al., 2000.
Oral	Adult with chronic renal failure (man, aged 59 years old)	Antacids containing aluminium	The total amount of ingested aluminium was at least 3 kg during 3 years	Encephalopathy, bone disease and microcytic anaemia (speech disturbances, mostly stuttering, hypochromic anaemia, hyperkalemia; the patient became confused and lethargic; the neurological disturbances progressed steadily, repeated general seizures occurred and the patient died).	Zatta et al., 2004.
Oral	Adult (woman, aged 44 years old)	Aluminium sulphate (present as flocculant in potable water supplies, accidentally released in high amounts)	500-3000 x the acceptable limit under European Union Legislation (0.200 mg/litre)/chronic (15 years after the accidental contamination of drinking water in Camelford)	Sporadic early-onset amyloid- β angiopathy- -Alzheimer's-related disease (difficulty in finding words, progressive dementia, visual hallucinations, headache, anxiety, cerebral ischemia, and death).	Exley and Esiri, 2006.

Exposure pathway	Population/Sample	Aluminium source and/or compound	Dose and/or duration	Aluminium effects	Reference
Intravenous	Dialysis patients (aged 15–61 years old at the start of the dialysis treatment)	Aluminium-containing dialysis fluid (derived from tap water treated with aluminium sulphate)	1 ppm/chronic (2-5 years)	Speech impairments (stuttering, dysarthria, dyspraxia, and motor aphasia), movement disorders (twitches, tremors, myoclonic jerks, seizures, and motor apraxia), cognitive impairments and behavioural changes (progressive dementia, paranoia, confusion, and psychosis), and death.	Flendrig et al., 1976.
Intramuscular injection	Macrophagic myofasciitis (MMF) syndrome patients (mean age 45 years old)	Aluminium hydroxide as a vaccine adjuvant	1–17 doses of aluminium-containing vaccines (hepatitis B, hepatitis A, and tetanus toxoid) in the period of 10 years prior to disease diagnosis	Myalgia, arthralgia, chronic fatigue (disabling fatigue >6 months), muscle weakness and cognitive dysfunction (overt cognitive alterations affecting memory, and attention). Patients develop an autoimmune disease, most frequently being multiple sclerosis-like demyelinating disorders, Hashimoto's thyroiditis, and diffuse dysimmune neuromuscular diseases, even in the absence of overt autoimmune disease, low titres of autoantibodies, increased inflammatory biomarkers.	Gherardi et al., 1998. Ryan et al., 2006. Couette et al., 2009. Passeri et al., 2011. Gherardi and Authier, 2012.
Inhalation	Adults (mean age 39 years old)	Aluminium oxide fumes during occupational exposure	0.13-1.95 mg/m^3/chronic	Headache, emotional irritability, concentration difficulty, insomnia, mood lability.	Sinczuk-Walczak et al., 2003.
Cerebrospinal fluid leakage	Adult (woman aged 52 years old)	Aluminium containing cement of bone reconstruction	30 mg/subacute (6 weeks)	Disturbance of consciousness, subacute coma, and grand mal seizures.	Reusche et al., 2001.

A correlation between chronic exposure to aluminium in the drinking water and the increased incidence of Alzheimer's disease was found in epidemiological studies (Rondeau et al., 2000). Aluminium interferes with the levels of amyloid-β peptides. Both *in vivo* and *in vitro* analyses show that aluminium does not affect the processing or expression of amyloid precursor protein (Neill et al., 1996).

However, the metal causes aggregation of amyloid-β peptides (Kawahara et al., 1994) and aggravates oxidative stress, amyloid-β deposition, and senile plaque formation in the brain of transgenic mice that overexpress the amyloid precursor protein (Praticò et al., 2002). Both amyloid-β and aluminium are able to stabilize the ferrous form of iron (Yang et al., 1999) and thus the two compounds can potentiate the formation of reactive oxygen species. The physical properties of aluminium ion allow its substitution in enzymes and structural proteins for the essential metals, magnesium and ferric iron (Shaw et al., 2014). The outcomes of the synergism between aluminium ion and ferrous iron are significantly higher peroxide levels in cells, peroxidative damage and oxidative stress (Ferrari, 2012; Kumar and Gill, 2014).

Aluminium also causes aggregation of microtubule-associated protein tau to a product resistant to proteases and phosphatases (Li et al., 1998). Aluminium interaction with hyperphosphorylated tau protein causes the tau to aggregate into granules of increasing size that eventually fuse to form cytoplasmic pools in human neurons of Alzheimer's brains (Walton, 2014). In brains of dialysis patients, it was observed that depletion of normal tau and accumulation of hyperphosphorylated tau correlated with brain aluminium levels. In fact, the tau changes significantly increase with rising levels of brain aluminium as well as amyloid-β accumulation (Harrington et al., 1994).

Intensification of neuroinflammation is seen in Alzheimer's disease patients (Colangelo et al., 2002). Coincidently, aluminium enhances neuroinflammatory events in the brain. Aluminium's pro-oxidative properties contribute to neuroinflammation by upregulating genes that have promoters enriched in binding sites for stress-inducible transcription factors; namely, the nuclear factor for kappa B (NFκB) and hypoxia inducible factor (HIF-1) (Lukiw et al., 2005). Intracerebroventricular injection of aluminium salts leads to activation of glial cells in the rat brain (Platt et al., 2001). Aluminium increases the production of the tumour necrosis factor alpha (TNFα) and promotes activation of the nuclear factor for kappa B in human glioblastoma cell lines (Campbell et al., 2002). In mice treated for ten weeks with aluminium lactate in the drinking water, there was a dose-dependent increase in pro-inflammatory cytokines (Campbell et al., 2004). In these animals, the highest number of astrocytes and activated microglia was detected in striatum.

Marked gliosis in the cortex is a prominent feature of Alzheimer's disease neuroinflammation (Beach et al., 1989). Gliosis is marked by elevated levels of the glial fibrillary acidic protein (GFAP) in reactive astrocytes. The Alzheimer's diseased brain is associated with activated microglia, reactive GFAP-producing astrocytes, and pro-inflammatory cytokines (Hoozemans et al., 2011). Gene microarrays of normal human neural cells, exposed to aluminium in primary cultures, showed upregulated gene expression for nuclear factor for kappa B, abnormal platelet amyloid-β protein precursor, cytokines, and other inflammatory proteins (Lukiw et al., 2005). In fact, 75% of pro-inflammatory and pro-apoptotic genes, that are either upregulated or down-regulated by a factor of 3-fold or more in Alzheimer's diseased brain tissue (Colangelo et al., 2002) are also altered in aluminium-exposed human neural cells (Lukiw et al., 2005).

The term "autism spectrum disorders" describes a variety of brain disorders that appear in infants or young children. Autism is usually characterized by delays in speech development and social functioning. Correlation between the number of aluminium-adjuvanted vaccines and autism spectrum disorders has been observed (Shaw et al., 2014). There is substantial evidence that the blood brain barrier is actually disrupted in autism and that this disruption, thought to be caused by environmental inflammatory stress triggers, leads to neuroinflammation and autoimmunity. Aluminium is known to damage the blood brain barrier and can increase its permeability by increasing the rate of transmembrane diffusion and by selectively altering saturable transport systems (Yokel, 2006). The breakdown of the blood brain barrier by aluminium may also result from excessive release of pro-inflammatory cytokines from aluminium-stimulated microglia. The ability of aluminium adjuvants to cross the blood brain barrier (Gherardi and Authier, 2012) and up-regulate chemoattractants, such as the monocyte chemoattractant protein-1 (MCP-1) (Seubert et al., 2008), could promote active recruitment of immunocompetent cells to the brain, leading to both widespread autoimmunity and deleterious inflammatory processes. Recently, a new syndrome was described: "autoimmune/inflammatory syndrome induced by adjuvants" (Guimarães et al., 2015). Aluminium adjuvants exert their immunostimulatory effect through many different pathways that activate both the innate and adaptive immune systems. One of the most significant is the activation of the NLRP3 inflammasome pathway. NLPR3 activation has been shown to trigger type 2 diabetes. By using NLPR3 knockout mice it has been demonstrated that the absence of inflammasome components leads to a better maintenance of glucose homeostasis and higher insulin sensitivity (Wen et al., 2011). On the other hand, activation of the inflammasome and its downstream components, pro-inflammatory cytokines IL-1β and IL-18, are strongly implicated in the development of several central nervous system disorders (Chakraborty et al., 2010).

Although the link between aluminium and the etiopathogenesis of neurodegenerative diseases remains controversial, the new lines of evidence make it difficult to deny the link. Moreover, it is widely accepted that aluminium is a neurotoxicant, and cause cognitive deficiency and dementia when it enters the brain. In particular, aluminium can affect children, elderly people, and patients with diminished renal functions.

Aluminium Interferes with Gaseous Neurotransmission

Nitric oxide (NO), carbon monoxide (CO) and hydrogen sulphide (H_2S) are the core gaseous neurotransmitters. In contrast to other neurotransmitters, gasotransmitters freely diffuse across cellular barriers and have a short biological half-life of seconds to few hours. Thereby, activation and inhibition of endogenous enzymatic production is crucial for acting as signalling molecules.

Aluminium and H_2S Signalling

H_2S is produced from L-cysteine by cystathionine-gamma-lyase (CSE, EC 4.4.1.1), cystathionine-beta-synthase (CBS, EC 4.2.1.22) and cysteine aminotransferase (CAT)/3-

mercaptopyruvate sulfurtransferase (3-MST, EC 2.8.1.2). It is also formed from D-cysteine by sequential action of D-amino acid oxidase (DAO, EC 1.4.3.3) and 3-MST. These enzymes show different levels of expression among mammalian tissues and assure the production of H_2S by four distinct pathways (Kimura, 2013). CSE and CBS are greatly expressed in kidney and liver and less in brain (Ishii et al., 2004). In the brain, H_2S is mainly generated by the antioxidant protein 3-MST in mitochondria (Shibuya et al., 2009; Nagahara, 2013).

CBS is preferentially expressed in the radial glial/astrocyte lineage (Enokido et al., 2005). This heme requiring enzyme is a risk factor for Alzheimer's disease, the upregulation of which in reactive astrocytes is associated with neuronal recovery from oxidative stress and ischemia–reperfusion injury (Kimura et al., 2010). Enhancement of antioxidant activity by H_2S relieves aluminium-induced stress in plant cells. In contrast to CBS, the aluminium resistant protein CSE is detected in neurons and activated in response to calcium/calmodulin (Yang et al., 2008). Cytosolic levels of CSE are low, but its activity in the human brain appears to be higher than in rodent brain (Diwakar and Ravindranath, 2007).

Activation of target proteins by sulfhydration is the most prominent signalling mechanism of H_2S (Kimura, 2014), which facilitates the induction of long-term potentiation in hippocampus, through regulation of N-methyl-D-aspartate (NMDA) receptors (Abe and Kimura, 1996). H_2S also stimulates calcium mobilization from internal stores (Bauer et al., 2010), and interferes with the activity of ATP-sensitive potassium channels and voltage-activated calcium channels, including facilitation of Cav3.2 in neurons (Fukami and Kawabata, 2015). In contrast to plant cells, where the capability of H_2S to reverse cellular injury in response to aluminium exposure is documented (Ali et al., 2015), the relevance of H_2S to aluminium toxicity in mammalians remains to be elucidated. H_2S is involved in modulating multiple physiological responses of relevance to aluminium neurotoxicity. H_2S is an anti-inflammatory signalling molecule that potentiates antioxidant response and exerts anti-apoptotic actions (Kimura, 2014). In experimental models of neurodegenerative diseases, H_2S appears to provide neuroprotection by preventing movement impairment associated with Parkinson's disease (Kida et al., 2011) and cognitive impairment associated with Alzheimer's disease (Xuan et al., 2012).

Aluminium and CO Synthesis

The inducible isoform (HO-1) and the constitutive isoform (HO-2) of heme oxygenase (HO, EC 1.14.99.3) generate CO, along with biliverdin/bilirubin and ferrous iron, from heme. Its release from heme-containing proteins is exacerbated during oxidative damage, and HO levels in the senescent and Alzheimer-diseased brain are increased (Schipper et al., 2000).

While biliverdin and bilirubin have mainly antioxidative properties that protect neurons from damage (Motterlini et al., 1996), iron ions catalyse Fenton and Haber-Weiss reactions that might enhance oxidative stress in aluminium-related neurodegenerative diseases. In response to oxidative stress, concomitant overexpression of calmodulin and HO-2, which is highly expressed in neurons (Ewing and Maines, 1992) and transiently activated by calcium/calmodulin during neuronal activity (Boehning et al., 2004), has been observed in injured brain (Alm et al., 2000). Under these conditions, in addition to CO production by neurons, glial production could contribute to elevate CO concentrations, given that the expression of HO-1 is induced in astrocytes, oligodendrocytes and microglia (Bidmon et al., 2001; Imuta et al., 2007). Apparently, HO-1 is also a component of the NO signalling cascade, the expression of which is triggered not only by the former gasotransmitter but also by a myriad of injuring stimuli (Motterlini et al., 1996).

Multiple alterations of the CO signalling pathway have been described in Alzheimer's disease, Parkinson's disease and amyotrophic lateral sclerosis (Wu and Wang, 2005). Besides the onset of aluminium adverse actions in the brain have been linked to neurodegeneration, aluminium effect on the CO-synthesizing enzymes was not addressed experimentally. However, the activity of these enzymes could be jeopardized through the capability of aluminium to interfere with iron and calcium binding proteins, such as ferritin and calmodulin (Shaw et al., 2014). Moreover, aluminium decreases the total antioxidant capacity (Ferrari, 2012), which favours the progression of protein and lipid peroxidation processes driven by iron and, consequently, intensifies neurodegeneration. Accordingly, aluminium-induced neurobehavioural and neurochemical changes are alleviated by antioxidants (Allagui et al., 2014).

In general, heme-containing proteins are putative targets of CO (Heinemann et al., 2014). Among them, nitric oxide synthase (NOS, EC 1.14.13.39) and the soluble forms of guanylate (guanylyl) cyclase (EC 4.6.1.2) and of adenylyl cyclase (EC 4.6.1.1), which generate the secondary messengers cGMP and cAMP, respectively, are of outstanding importance for cell signalling. In contrast to H_2S, CO and NO inhibit NOS and stimulate the activities of guanylyl and adenylyl cyclases (Stone and Marletta, 1994; Middelhaufe et al., 2012), which are common denominators in the signalling cascades of these gasotransmitters. Enhancement of oxidative stress through upregulation of HO and NOS in injured brain accounts for many neurodegenerative conditions (Alm et al., 2000).

Aluminium and Nitroxidergic Neurons

NO is formed by a family of isoforms of the heme requiring NOS, namely neuronal NOS (nNOS), endothelial NOS (eNOS), and inducible NOS (iNOS). The constitutively expressed isoforms, nNOS and eNOS, are activated by calcium/calmodulin (Bredt and Snyder, 1990).

The association of NO signalling to aluminium toxicity was one of the firsts to be biochemically established. Activation of the NO system was postulated as the main mechanism of gastroprotection by aluminium-containing antacids (Konturek et al., 1992). More recently, alterations of NO signalling are recurrently associated with aluminium-induced brain injury. Long lasting exposure to aluminium reduces the expression of nNOS and even the number of nitroxidergic neurons in brain cortex (Rodella et al., 2006). More frequent is the observation that NOS activity in brain is raised up during exposure to aluminium with concomitant enhancement of NO levels (Stevanović et al., 2009a). In the brain, aluminium is capable of enhancing iNOS expression by phagocytic and immunocompetent cells, which produce pro-inflammatory cytokines in noteworthy amounts (Prakash et al., 2013; Zaky et al., 2013). Probably, the inducible form of NOS mediates the pro-oxidant actions of aluminium (Stevanović et al., 2009b), which in turn result on disseminated neuronal death, being neuropeptide Y/nitroxidergic and cholinergic pathways among the most affected (Bielarczyk et al., 2003; Rodella et al., 2006; Stevanović et al., 2011). Aluminium-induced apoptotic cell death was effectively prevented by pre-treatment of PC12 cells with iNOS inhibitors, while nNOS inhibitors had proved ineffective (Satoh et al.,

2007). During intrauterine developmental stages, the neuronal NO signalling pathway is also targeted by aluminium, causing loss of nitroxidergic neurons (Kim, 2003).

NO might activate or inhibit target proteins straightly by S-nitrosylation of SH groups of cysteines (Stamler et al., 1992). NO binding to heme in the active site of guanylyl cyclase, promoting the formation of cGMP, is still considered the main mechanism of its action as signalling molecule. Aluminium causes lessening of NO-induced activation of guanylyl cyclase (Llansola et al., 1999; Canales et al., 2001). In cerebellar neurons, this effect is accompanied by the reduction of glutamate-induced activation of NO synthase and cell death, most probably through the N-methyl-D-aspartate (NMDA) receptor-associated signal transduction pathway (Llansola et al., 1999). Thereby, impairment of the neuronal glutamate-NO-cGMP pathway has been referred as one of most plausible mechanisms of aluminium-induced cell death (Cucarella et al., 1998; Hermenegildo et al., 1999) and decline of hippocampal long-term potentiation (Platt et al., 1995). However, similarly to other neurodegenerative disorders, NO-mediated decline of cortical cholinergic neurotransmission might also contribute to the reduced short-term memory capacity in intoxication by aluminium (Szutowicz, 2001; Stevanović et al., 2011).

Reversal of Plasma Membrane Neurotransmitter Transporters Is Dysregulated by Aluminium

Similarly to gasotransmitters, the release of non gaseous ones also depends on their availability, due to the performance of the respective metabolic pathways. Beside synthesis and cleavage of neurotransmitters, which is out of the scope of this chapter, we would like to note that significant alterations of neurotransmitter levels in aluminium overload are considered to be region-specific and dependent on the dose and duration of exposure (Lai et al., 1982; Sturman et al., 1983; Houser et al., 1986; Beal et al., 1989; Julka et al., 1995; Tsunoda and Sharma, 1999; Struys-Ponsar et al., 2000; Nayak and Chatterjee, 2001; Kumar, 2002).

Aluminium and Neurotransmitter Transporters

Recently, three neurotransmitter transporters (EAT-4, MOD-5 and DAT-1) have been suggested as molecular targets of Al_2O_3 nanoparticles in the nematode worm *Caenorhabditis elegans* with impaired locomotion behaviour (Li et al., 2013). Aside from the EAT-4, of which the amino acid sequence homology with the human vesicular glutamate transporters is high, MOD-5 and DAT-1 are orthologous to the human plasma membrane neurotransmitter transporters (PNTs) of 5-hydroxytryptamine SLC6A4 and of dopamine/noradrenaline SLC6A2. SLC6 stands for solute carrier 6 gene family, also referred to as the neurotransmitter-sodium-symporter family, which comprises a group of nine sodium- and chloride-dependent PNTs for γ-aminobutyric acid (GABA), glycine, 5-hydroxytryptamine, dopamine and noradrenaline.

As far as we know, there is no experimental demonstration that PNTs are primary targets for aluminium action in mammalian brain. Meanwhile, modification of neurotransmitter release has been recurrently suggested to be involved in the deleterious effects of aluminium on brain circuits mediated by acetylcholine, 5-hydroxytryptamine, catecholamines, GABA and glutamate (Banin and Meiri, 1987; Beal et al., 1989; Albrecht and Norenberg, 1991; Albrecht et al., 1991; Zielke et al., 1993; Gilbert and Shafer, 1996; Bielarczyk et al., 1998; Szutowicz et al., 1998; Struys-Ponsar et al., 2000; Amador et al., 2001; Cordeiro et al., 2003; Silva et al., 2007; Laabbar et al., 2014). Aside from acetylcholine, which is cleaved by acetylcholinesterase in the synaptic cleft, the presynaptic homeostasis of the above mentioned neurotransmitters is contributed by PNTs. This contribution is even more acute when releasing pools are particularly reduced in the absence of reuptake, such as in the cases of dopamine, noradrenaline and 5-hydroxytryptamine.

The aluminium-induced learning and memory deficit suggests that it includes changes in neurotransmitter levels other than acetylcholine and glutamate. For instance, impaired learning and memory ability in aluminium exposed rats has been associated with high levels of GABA in neocortex (Shuchang et al., 2008), increased 5-hydroxytryptamine levels in olfactory lobe, hippocampus, striatum, hypothalamus, midbrain, pons, medulla oblongata and spinal cord, whereas its levels in cortex, hippocampus and cerebellum were decreased (Kumar, 2002).

Aluminium Effect on Sodium-Dependent Transport of Neurotransmitters

Most of the studies addressing the effect of aluminium on neurotransmitter release do not provide a clear distinction between PNTs-mediated release and soluble *N*-ethylmaleimide-sensitive factor attachment protein receptor (SNARE)-mediated, calcium-dependent exocytosis, which will be the topic of the next section.

There are very few studies focusing into the effect of aluminium on neurotransmitter release by the reversal of PNTs. The PNT-mediated GABA release by isolated presynaptic nerve terminals is slightly enhanced by micromolar aluminium chloride (Cordeiro et al., 2003), which causes significant stimulation of the release of GABA and glutamate from cultured cortical astrocytes (Albrecht and Norenberg, 1991; Albrecht et al., 1991). Alteration of the glutamate–glutamine pathway, supported by astrocytes and neurons, appears to characterize brain regions showing high aluminium levels and also in patients with dialysis encephalopathy (Zielke et al., 1993; Struys-Ponsar et al., 2000). Disturbance of the GABAergic system is a feature of dialysis encephalopathy, given that its levels are reduced in specific regions (frontal, occipital, and cerebellar cortex, caudate nucleus, and medial dorsal thalamus) of autopsied brains of patients (Sweeney et al., 1985). Glial cells and GABAergic neurons are more prompt to take up aluminium (De Stasio et al., 1994; Aremu and Meshitsuka, 2005), and consequently more vulnerable to aluminium-induced insult. Aluminium hydroxide gel/alumina cream is used to cause selective loss of GABAergic neurons in animal models of epilepsy (Houser et al., 1986). The efficacy of GABA analogues (e.g., gabapentin) is high for treating tonic-clonic/myoclonic seizures, a clinical feature of aluminium encephalopathy. Meanwhile, acetylcholinesterase inhibitors and *N*-methyl-D-aspartate (NMDA) receptor antagonists still provide the best palliative relief of symptoms in sporadic forms of Alzheimer's disease (Nelson and Tabet, 2015).

In contrast to the few studies showing aluminium-induced enhancement of PNTs-mediated release of neurotransmitters, the inhibition of their uptake back to presynaptic nerve terminals has been reported repeatedly. In synaptosomal preparations of rabbit hippocampus and spinal cord, in which chronic neurofibrillary changes had been induced with aluminium chloride, the high-affinity uptake of GABA and glycine was found to be reduced (Sturman et al., 1983). The *in vitro* effect of aluminium on the uptake of noradrenaline, 5-hydroxytryptamine, glutamate and GABA was contrasted in rat brain synaptosomal fractions (Wong et al., 1981; Lai et al., 1982). Sub-millimolar concentrations of aluminium are required to sharply reduce the amount of neurotransmitters taken up by synaptosomes. At these concentrations, aluminium also inhibits the synaptosomal Na/K-ATPase (Cordeiro et al., 2003; Silva et al., 2007). Thereby, dissipation of the sodium-gradient across the plasma membrane should account for the inhibitory effect on sodium-dependent transport of neurotransmitters. It is important to note that the Na/K-ATPase activity is significantly decreased in aluminium intoxicated brain (Silva et al., 2005, 2013; Bihaqi et al., 2009; Sumathi et al., 2014).

Aluminium interferes with the translocation of neurotransmitters across the plasma membranes by PNTs differently, when mediating the release or the uptake of amino acids and monoamines. The localization, specificity, affinity, capacity, coupling to the energy-dependent transport of sodium ions and sensitivity to modulators of PNTs of amino acid and monoamine neurotransmitters largely contribute to regulate synaptic firing. Thereby, aluminium-induced reduction of the uptake of GABA (Wong et al., 1981; Sturman et al., 1983; Cordeiro et al., 2003), glutamate (Wong et al., 1981), glycine (Sturman et al., 1983), dopamine (Lai et al., 1981), noradrenaline and 5-hydroxytryptamine (Lai et al., 1982) in intact nerve terminals of mammalian brain would also contribute to depressed neurotransmission by disabling neurotransmitter recycling at the synapse.

Aluminium Effect on Modulators of Neurotransmitter Transporters

The release of neurotransmitters to the synaptic cleft and its reuptake into neurons and astrocytes are the foremost recognized functions of PNTs among their various functions in physiological, pathological and pharmacological conditions (Raiteri and Raiteri, 2015). PNTs also exhibit multiple conductance states that might impact synaptic excitability faithfully (Sung et al., 2003).

Phosphorylation-dephosphorylation changes the capability of PNTs for translocation of neurotransmitters (Steiner et al., 2008), regulates their trafficking between membranous compartments (Ramamoorthy et al., 2011) and modifies the interactions of PNTs with integral membrane proteins of the exocytotic machinery. Here, we underline the interaction of PNTs for dopamine, noradrenaline, 5-hydroxytryptamine and GABA with the target Q-SNARE protein syntaxin 1A (Beckman et al., 1998; Deken et al., 2000; Geerlings et al., 2000; Haase et al., 2001; Sung et al., 2003). This integral protein of the plasma membrane, involved in the calcium-regulated exocytosis of neurotransmitter vesicular pools, is engaged in the regulation of neurotransmitter clearance from the synaptic cleft by modulating the activity and trafficking of PNTs. Insertion of PNTs into plasma membrane is favoured by syntaxin 1A. Esters of phorbol, through protein kinase C (PKC, EC 2.7.11.13), induce dissociation of syntaxin 1A from PNTs, and consequent failure of transporter-mediated translocation of dopamine, noradrenaline, 5-hydroxytryptamine and GABA. Dissociation of syntaxin 1A from PNTs is also promoted by okadaic acid, an inhibitor of protein-serine/threonine phosphatase (PP, EC 3.1.3.16) activity, especially of PP1 and of the PNTs associated PP2A.

Aluminium action on neurotransmitter translocation across presynaptic plasma membranes through regulation of PNTs by phosphorylation-dephosphorylation was challenged in GABAergic neurons. Aluminium does not change the rate of GABA homoexchange, but completely reverses the calcium inhibitory action on its uptake and release (Cordeiro et al., 2003). Phosphorylation of internally localized calcium/calmodulin-dependent protein phosphatase 2B (calcineurin/PP2B)-sensitive sites of the transporter appears to assure uptake and release of GABA at maximal rates (Gonçalves et al., 1999; Cordeiro et al., 2000). Calcium/calmodulin, through calcium/calmodulin-dependent protein kinase II (CaMKII, EC 2.7.11.17), also favours the reversal of dopamine transporter (Fog et al., 2006). Incubation of synaptic plasma membrane fraction isolated from brain cortex with aluminium completely prevents stimulation of calcineurin by calcium/calmodulin (Cordeiro et al., 2003), suggesting that the phosphorylation state of PNTs might be disturbed in overloaded brain with aluminium.

It has been proposed that decreased neurotransmission in Alzheimer's disease could be in part attributed to enhancement of calcineurin activity in aged brain, which is further activated by oligomers of the 42-residue amyloid-β peptide, the major constituent of senile plaques in Alzheimer's disease (Yu et al., 2009; Reese and Taglialatela, 2011). It is reasonable to wonder whether amyloid-β oligomers and aluminium that shift the regulation of PNTs by calcineurin in opposite directions both assure silencing of synapses during early states of neurodegeneration. However, complexes formed by the 42-residue peptide amyloid-β and aluminium were shown to selectively change the expression of genes involved in glutamatergic transmission, calcium homeostasis, oxidative stress, inflammation, and neuronal apoptosis in human neuroblastoma cells (Gatta et al., 2011). Dysregulation of calcium dynamics, which is thoroughly related with low levels of synaptic plasticity and excitotoxicity, metallic ion overload, which potentiate oxidative stress, and even accumulation of amyloid-β peptide take part in the regular process of brain aging (Dewachter et al., 2000).

Aluminium Effect on Na/K-ATPase and Neurotransmitter Release

According to the amyloid hypothesis, Alzheimer's disease relies on abnormal accumulation of amyloid-β peptide, the principal physiological function of which appears to be to depress synaptic activity, counteracting the action of either other proteolytic products of its precursor (amyloid precursor protein) or of the amyloid precursor protein itself, in order to avoid excitotoxicity (Pearson and Peers, 2006; Kawahara and Kato-Negishi, 2011). Accumulation of amyloid-β peptides and metals (aluminium, copper, iron and zinc) in Alzheimer's brain areas has been reported in the literature (Exley et al., 2012). Most probably, both aluminium and amyloid-β peptide would exhaust synapses synergistically by inhibiting the Na/K-ATPase.

In human brain, amyloid-β peptides can form highly toxic 10-15-nm spherical assemblies, known as amylospheroids (Hoshi et al., 2003). Recent works on molecular mechanisms of Alzheimer's disease drew attention to amylospheroids, given that they induce cell death of functionally mature hippocampal neurons by selective binding to the α3-isoform of the Na/K-ATPase catalytic subunit (Ohnishi et al., 2015). Restoration of the resting membrane potential subsequent to depolarization and repeated firing of action potentials is mainly assured by maximal activation of the α3-isoform. Amylospheroids not only inhibit α3 activity selectively but also downregulate its expression. The decrease of expression level of one of the Na/K-ATPase α-subunit isoforms is usually associated with a shift from one isoform to another or accompanied by a parallel decrease of all isoform levels (Charlemagne et al., 1994; Serluca et al., 2001). *In vivo* aluminium overload of rat brain does not interfere with the expression of Na/K-ATPase isoforms but inhibits their activity (Silva and Gonçalves, 2003; Silva et al., 2005). Thereby, aluminium might exacerbate the inhibitory effect of amylospheroids on neuronal Na/K-ATPase. Ouabain, a selective inhibitor of Na/K-ATPase, was shown to reproduce the main morphological features of amylospheroids-induced cell death (Ohnishi et al., 2015).

We wonder whether amylospheroids also act as secretagogues as ouabain does (Levi and Raiteri, 1993). Apparently, similar action mechanisms assure cell death induced by amylospheroids and neurotransmitter release induced by ouabain: the failure in active transport of sodium and potassium ions by inhibition of Na/K-ATPase causes the enhancement of cytosolic sodium levels, membrane depolarization and elevation of cytosolic calcium levels, which in turn favour the release of neurotransmitters.

Aluminium Reduces SNARE-Mediated, Calcium-Dependent Exocytosis

SNARE-mediated, calcium-dependent exocytosis or regulated exocytosis refers to the most ubiquitous mechanism of release of neurotransmitters packed in synaptic vesicles. Its dependency on elevation of intracellular calcium concentrations stands as a property of this secretory route.

Aluminium Effect on Presynaptic Calcium and Neurosecretion

Aluminium overload causes dysregulation of calcium homeostasis that resembles those observed in aged and Alzheimer's diseased brains (Walton, 2012). Elevation of free calcium concentration in presynaptic nerve terminals exposed to aluminium has been documented in the literature and several action mechanisms advanced: inhibition of calcium extrusion by the plasma membrane Ca-ATPase (Kaur and Gill, 2005); enhancement of calcium influx through voltage-operated and neurotransmitter-gated calcium channels and through the plasma membrane Na/Ca antiporter (Mundy et al., 1997; Kaur and Gill, 2005); and mobilization of intracellular calcium stores (Kodavanti et al., 1993). Apparently, elevated intracellular calcium concentrations should support exocytotic release of neurotransmitters in response to stimulation. Accordingly, a brief increase of spontaneous and stimulation-induced release of acetylcholine upon aluminium addition, at concentrations similar to those found in the diseased brain of demented patients, has been observed at the frog neuromuscular junction; yet, irreversible discontinuation of releasing activity rapidly replaced the initial stimulatory effect of aluminium (Banin and Meiri, 1987). As far as we know, there are no reports on aluminium induced enhancement of SNARE-mediated, calcium-dependent exocytosis in mammalian brain.

Generally, decreased neurotransmission appears to precede other negative outcomes (synaptic loss, neuroinflammation and cell death), observed in aluminium intoxicated brain and in other pathological conditions in which neurodegeneration constitutes the most prominent feature at cellular level. Accordingly, we observed a reduction of acetylcholine release by rat brain cortical presynaptic nerve terminals when perfused with micromolar aluminium (Silva et al., 2007). Aluminium doses that cause cognitive deficit in rats also inhibit the calcium-dependent, potassium-evoked glutamate release in hippocampal slices (Yokel et al., 1987; Provan and Yokel, 1992). From those studies it is risky to conclude about the specific input of SNARE-mediated, calcium-dependent exocytosis to the total amount of acetylcholine and glutamate released, given that for both neurotransmitters alternative release mechanisms are also activated in response to stimulation, namely mediatophore-mediated acetylcholine release (Israël et al., 1988; Dunant, 2006) and efflux of glutamate through PNTs (Rossi et al., 2000).

Aluminium and the Fusion Pore Properties

SNARE-mediated, calcium-dependent exocytosis is a multistage process, which is divided in three main steps: docking, priming and fusion of synaptic vesicles. The last step, the merger of the synaptic vesicle membrane with the plasma membrane, is calcium-dependent and requires the preassembly of a *trans*-complex between three SNARE proteins, one localized in the synaptic vesicle membrane (synaptobrevin/VAMP) and the others (syntaxin 1A and SNAP 25) in the plasma membrane. The transition of the SNARE complex from the '*trans*' to the '*cis*' conformation coincides with the beginning of the establishment of the fusion pore, through which the lumen of the synaptic vesicle is exposed to the synaptic cleft. The fusion pore starts as hemi-fusion stalk that is rapidly transformed into a narrow pore, the dilation of which allows the release of vesicular cargo, i.e., the neurotransmitter release. This process is assisted by proteins and lipids, such as Munc 18-1 and cholesterol (Brunger, 2005; Rituper et al., 2012). The fusion pore is a dynamic structure; it can either retract to the narrow conformation during kiss-and-run exocytosis and pore flickering or, alternatively, further dilate till complete merger of the two cognate membranes during full fusion (Vardjan et al., 2013).

The precise question whether aluminium interferes with SNARE-mediated, calcium-dependent exocytosis was addressed in rat pituitary lactotrophs in culture (Calejo et al., 2012; Guček et al., 2015). SNARE-mediated, calcium-dependent exocytosis is the only mechanism supporting the release of prolactin by lactotrophs. Micromolar aluminium was shown to reduce prolactin secretion by decreasing the frequency of unitary exocytotic events and by stabilizing the diameter of the majority of the fusion pores to a value sufficiently narrow to prevent the outflow of prolactin molecules. Reduction of peptide secretion frequency and of vesicle mobility, as well as dysregulation of cytoplasmic calcium homeostasis have been shown in astrocytes expressing a mutated form of presenilin 1 related to Alzheimer's disease (Stenovec et al., 2016). Enhancement of presenilin 1 expression by *in vivo* exposure to aluminium has been observed in brain cortex and hippocampus (Wang et al., 2014). Thereby, the effect of aluminium on synaptic vesicle mobility deserves attention. The importance of astrocytic vesicle traffic for neurotransmission and neurodegeneration was recently highlighted by Vardjan et al. (2015). Exposure to aluminium also disabled the prolongation of the fusion pore lifetime induced by stimulation, yet aluminium caused a 3-fold enhancement of this parameter during non-stimulating conditions. Although the exact mechanism underlying the aluminium-induced reduction of prolactin secretion was not completely elucidated, it was clearly shown that aluminium interferes with the ultimate step of SNARE-mediated, calcium-dependent exocytosis, the dilation of the fusion pores. We wonder whether fusion pores are also more stable and narrower when aluminium is in the synapse.

While aluminium effect on the fusion pore properties was only depicted in prolactin secreting cells, it could be interpreted as indicative of its capability to promote general neurotransmission suppression, rather than an evidence of selective inhibition of neuropeptide exocytosis. The diameter of the fusion pore, formed as a result of SNARE-ring complexes self-assemble, appears to change with the size of the fused synaptic vesicle. Small SNARE-ring complexes (Ø ≈ 20 nm) were generated by using small proteoliposomes (Ø ≈ 50 nm), the diameter of which linearly increased when larger proteoliposomes (Ø $\leq$ 800 nm) were employed (Cho et al., 2005). In mammalian brain, synaptic vesicles are heterogeneous populations in packed neurotransmitters, density and size; their diameters vary from less 50 nm (small clear synaptic vesicles) to more than 200 nm (large dense-core vesicles), establishing fusion pores ranging from ~0.6 nm to ~2.0 nm, respectively, during transient fusion at rest conditions (Klyachko and Jackson, 2002). We venture to suggest that aluminium might reduce the responsiveness of nerve terminals to stimulation by limiting the amount of secreted neurotransmitters, due to diminution of exocytotic events, shortening of the lifetime and conductance of fusion pores. A facilitation of secretion of low-molecular weight neurotransmitter during resting conditions by aluminium should not be discarded, given that it seems to delay the closure of fusion pores in the absence of stimulation (Calejo et al., 2012). The physiological relevance of constrained diffusion of vesicular cargo through fusion pores because of close match between their diameter and the size of neurotransmitter molecules, as well as of copackage of low-molecular weight neurotransmitters and neuropeptides together have been highlighted during the past decades (Hu et al., 2008; Zhang et al., 2011; Flašker et al., 2013). How fusion pore widens/narrows is regulated at molecular level remains vague (Jorgačevski et al., 2012; Guček et al., 2015; Hammel and Meilijson, 2016). Herein we draw attention to the critical role of cholesterol in SNARE-mediated, calcium-dependent exocytosis. Although SNARE protein clusters are not localized in plasma membrane rafts enriched in cholesterol, its removal disintegrates SNARE protein clusters and reduces calcium regulated exocytosis (Lang et al., 2001). The effect of cholesterol depletion and repletion by methyl-β-cyclodextrin in lactotrophs has been used as evidence for its role in unitary exocytotic events. A 3.5-fold drop of cholesterol content reduces the frequency of exocytotic events by 2.5 times and the fusion pore lifetime during stimulation, most probably due to dispersal of SNARE proteins and net removal of negatively curved lipids (Rituper et al., 2012). It has been suggested that the onset of cognitive decline in neuropathologies such as Alzheimer's, Huntington's and Parkinson diseases could be influenced by impaired regulation of neurotransmission related to cholesterol homeostasis failure in the central nervous system (Segatto et al., 2014). Dietary aluminium might cause significant decrease of the cholesterol content of presynaptic nerve terminals and concomitant reduction of membrane cholesterol/phospholipid molar ratio (Silva et al., 2002, 2013). Moreover, to the stabilization of the tubular curvature of the fusion pores in the synaptic membrane environment, the contribution of anisotropic constituents with high negative curvature and cations adjacent to the fusion pore is likely important (Jorgačevski et al., 2010; Kabaso et al., 2012; Calejo et al., 2014; Guček et al., 2015). Membrane cholesterol and phosphatidylethanolamine levels both decline after long-lasting exposure to aluminium and to the hypolipidemic agent colestipol (Silva et al., 2002, 2013).

Although the exact mechanism underlying the aluminium-induced reduction of SNARE-mediated, calcium-dependent exocytosis is not yet completely elucidated, it is plausible to assume that in aluminium overload the function of fusion pores is jeopardized, meaning that aluminium also comes in opposition with neurosecretion downstream of the stimulation-secretion coupling. Interestingly, monomeric and oligomeric forms of amyloid-β peptide bind specifically to the SNARE motif region of syntaxin 1A, allowing synaptic vesicles to dock but preventing the formation of the fusion pore (Yang et al., 2015).

We wonder if aluminium could depress calcium regulated exocytosis through either calcineurin or *N*-ethylmaleimide sensitive factor. Calcineurin is thought to be a negative regulator of exocytosis, since enhanced exocytosis of glutamate by its inhibition in isolated mammalian nerve terminals was demonstrated several decades ago (Nichols et al., 1994). Aluminium hinders the activation of membrane-bound calcineurin in presynaptic membranes purified from rat brain cortex (Cordeiro et al., 2003), rendering improbable the involvement of calcineurin in the aluminium-induced decrease of neurotransmitter secretion. *N*-ethylmaleimide sensitive factor, the vesicle-fusing ATPase (EC 3.6.4.6) that disassembles inactive *cis*-SNARE complexes, is an oligomeric ATP/ADP-binding enzyme with low intrinsic ATPase activity (Zhao et al., 2015). There is a large body of evidence that aluminium inhibits adenosinetriphosphatase activity. Moreover, in mammalian brain overloaded with aluminium, it appears to drastically reduce the activation of Na/K-ATPase by high ATP concentrations through disassembly of the oligomeric structure of the membrane bound enzyme (Silva and Gonçalves, 2003; Silva et al., 2013). It is uncertain that aluminium will inhibit the vesicle-fusing ATPase activity, although it could also contribute to the reduced frequency of exocytotic events, by decreasing the probability of SNARE-mediated formation of fusion pores. On the contrary, evidence that dietary aluminium resulting on cholesterol and phosphatidylethanolamine diminution leads to decrease in lipid order of the hydrophilic and hydrophobic membrane regions in brain nerve terminals is supported by the fact that it decreases the steady-state fluorescence anisotropy of 1-(4-(trimethylamino)phenyl) -6-phenylhexa-1,3,5-triene (TMA-DPH) and 1,6-diphenyl-1,3,5-hexatriene (DPH) (Silva et al., 2002). In nerve terminals, synaptic vesicles fuse preferentially at the so-called active zones. Cholesterol depletion and aluminium might reduce calcium-dependent release of neurotransmitters by nerve terminals, as shown in isolated glutamatergic nerve terminals treated with methyl-β-cyclodextrin (Gil et al., 2005) and in hippocampal slices prepared from brains overloaded with aluminium (Yokel et al., 1987; Provan and Yokel, 1992).

CONCLUSION

Recent insights into the molecular basis of aluminium toxicity and neurodegenerative disorders, Alzheimer's disease in particular, amended our understanding of aluminium associated neuropathies. In neurodegeneration, neurotransmission decline, synaptic loss, neuroinflammation and cell death appear to be sequentially elicited during the progression of cognitive deficit. Aluminium-induced alterations of neurotransmitter release should account to neurotransmission decline.

A broad reduction of neurotransmitter release by SNARE-mediated, calcium-regulated exocytosis by aluminium is expectable, through alteration of the properties of the fusion pores, which formation is also prevented by the amyloid-β peptide. *In vivo* brain overload with aluminium hinders maximal activation of Na/K-ATPase, and amylospheroids bind to and inhibit the neuronal α3 isoform of the enzyme. Neurotransmitter recycling mediated by sodium-dependent PNTs is particularly affected by selective inhibition of Na/K-ATPase, which also triggers neurosecretion and cell death. Spreading of aluminium-induced neuronal loss seems to require free radical production, which is facilitated by activation of NOS and amyloid-β peptide. In conclusion, it is plausible to assume that environmental and health factors that exacerbate neurotransmission failure would speed the progression of aluminium-related neuropathies.

REFERENCES

Abe, Kazubo, and Hideo Kimura. 1996. "The possible role of hydrogen sulfide as an endogenous neuromodulator." *Journal of Neuroscience* 16:1066-71.

Albrecht, Jan, and Michael D. Norenberg. 1991. "Aluminum chloride stimulates NaCl-dependent release of taurine and γ-aminobutyric acid in rat cortical astrocytes." *Neurochemistry International* 18:125-9.

Albrecht, Jan, M. Simmons, G. R. Dutton, and M. D. Norenberg. 1991. "Aluminum chloride stimulates the release of endogenous glutamate, taurine and adenosine from cultured rat cortical astrocytes." *Neuroscience Letters* 127:105-7.

Alfrey, Allen C., Gary R. LeGendre, and William D. Kaehny. 1976. "The dialysis encephalopathy syndrome. Possible aluminum intoxication." *New England Journal Medicine* 294:184-8.

Ali, Basharat, P. Qian, R. Sun, M. A. Farooq, R. A. Gill, J. Wang, M. Azam, and W. Zhou. 2015. "Hydrogen sulfide alleviates the aluminum-induced changes in *Brassica napus* as revealed by physiochemical and ultrastructural study of plant." *Environmental Science and Pollution Research* 22:3068-81.

Alm, Per, H. S. Sharma, P. O. Sjöquist, and J. Westman. 2000. "A new antioxidant compound H-290/51 attenuates nitric oxide synthase and heme oxygenase expression following hyperthermic brain injury." *Amino acids* 19:383-94.

Altmann, Paul. 2001. "Aluminium induced disease in subjects with and without renal failure - Does it helps us understand the role of aluminium in Alzheimer's disease?" In: *Aluminum and Alzheimer's Disease. The Science that Describes the Link*, edited by Christopher Exley, 1-36. Amsterdam: Elsevier Science.

Altmann, Paul, J. Cunningham, U. Dhanesha, M. Ballard, J. Thompson, and F. Marsh. 1999. "Disturbance of cerebral function in people exposed to drinking water contaminated with aluminium sulphate: retrospective study of the Camelford water incident." *British Medical Journal* 319:807-11.

Amador, Fátima C., Maria S. Santos, and Catarina R. Oliveira. 2001. "Lipid peroxidation and aluminium effects on the cholinergic system in nerve terminals." *Neurotoxicity Research* 3:223-33.

Aremu, David A., and Shunsuke Meshitsuka. 2005. "Accumulation of aluminum by primary cultured astrocytes from aluminum amino acid complex and its apoptotic effect." *Brain Research* 1031:284-96.

Banin, Eyal, and Halina Meiri. 1987. "Impaired control of information transfer at an isolated synapse treated by aluminum: is it related to dementia?" *Brain Research* 423:359-63.

Bauer, Claudia C., J. P. Boyle, K. E. Porter, and C. Peers. 2010. "Modulation of Ca^{2+} signalling in human vascular endothelial cells by hydrogen sulfide." *Atherosclerosis* 209:374-80.

Beach, Thomas G., R. Walker, E. G. McGeer. 1989. "Patterns of gliosis in Alzheimer's disease and aging cerebrum." *Glia* 2: 420-436.

Beal, M. Flint, M. F. Mazurek, D. W. Ellison, N. W. Kowall, P. R. Solomon, and W. W. Pendlebury. 1989. "Neurochemical characteristics of aluminum-induced neurofibrillary degeneration in rabbits." *Neuroscience* 29:339-46.

Beckman, Matthew L., Eve M. Bernstein, and Michael W. Quick. 1998. "Protein kinase C regulates the interaction between a GABA transporter and syntaxin 1A." *The Journal of Neuroscience* 18:6103-12.

Bidmon, Hans-J., B. Emde, E. Oermann, R. Kubitz, O. W. Witte, and K. Zilles. 2001. "Heme oxygenase-1 (HSP-32) and heme oxygenase-2 induction in neurons and glial cells of cerebral regions and its relation to iron accumulation after focal cortical photothrombosis." *Experimental Neurology*, 168:1-22.

Bielarczyk, Hanna, A. Jankowska, B. Madziar, A. Matecki, A. Michno, and A. Szutowicz. 2003. "Differential toxicity of nitric oxide, aluminum, and amyloid β-peptide in SN56 cholinergic cells from mouse septum." *Neurochemistry International* 42:323-31.

Bielarczyk, Hanna, Maria Tomaszewicz, and Andrzej Szutowicz. 1998. "Effect of aluminum on acetylCoA and acetylcholine metabolism in nerve reminals." *Journal of Neurochemistry* 70:1175-81.

Bihaqi, Syed Waseem, M. Sharma, A. P. Singh, and M. Tiwari. 2009. "Neuroprotective role of *Convolvulus pluricaulis* on aluminium induced neurotoxicity in rat brain." *Journal of Ethnopharmacology* 124:409-15.

Blaylock, Russell L., and A. Strunecká. 2009. "Immune-glutamatergic dysfunction as a central mechanism of the autism spectrum disorders." *Current Medicinal Chemistry* 16:157-70.

Boehning, Darren, L. Sedaghat, T. W. Sedlak, and S. H. Snyder. 2004. "Heme oxygenase-2 is activated by calcium-calmodulin." *Journal of Biological Chemistry* 279:30927-30.

Bondy, Stephen C. 2014. "Prolonged exposure to low levels of aluminum leads to changes associated with brain aging and neurodegeneration." *Toxicology* 315:1-7.

Bowdler, Noelle C., D. S. Beasley, E. C. Fritze, A. M. Goulette, J. D. Hatton, J. Hession, D. L. Ostman et al., 1979. "Behavioral effects of aluminum ingestion on animal and human subjects." *Pharmacology, Biochemistry and Behavior* 10:505-12.

Bredt, David S., and Solomon H. Snyder. 1990. "Isolation of nitric oxide synthetase, a calmodulin-requiring enzyme." *Proceedings of the National Academy of Sciences* 87:682-5.

Brunger, Axel T. 2005. "Structure and function of SNARE and SNARE-interacting proteins." *Quarterly Reviews of Biophysics* 38:1-47.

Calejo, Ana I., J. Jorgačevski, B. Rituper, A. Guček, P. M. Pereira, M. A. Santos, M. Potokar et al. 2014. "Hyperpolarization-Activated Cyclic Nucleotide-Gated Channels and cAMP-Dependent Modulation of Exocytosis in Cultured Rat Lactotrophs." *The Journal of Neuroscience* 34:15638-47.

Calejo, Ana I., J. Jorgačevski, V. S. Silva, M. Stenovec, M. Kreft, P. P. Gonçalves, and R. Zorec. 2012. "Aluminium-induced changes of fusion pore properties attenuate prolactin secretion in rat pituitary lactotrophs." *Neuroscience* 201:57-66.

Campbell, Arezoo, A. Becaria, D. K. Lahiri, K. Sharman, and S. C. Bondy. 2004. "Chronic exposure to aluminum in drinking water increases inflammatory parameters selectively in the brain." *Journal of Neuroscience Research* 75:565-72.

Campbell, Arezoo, E. Y. Yang, M. Tsai-Turton, and S. C. Bondy. 2002. "Pro-inflammatory effects of aluminum in human glioblastoma cells." *Brain Research* 933:60-5.

Canales, Juan J., R. Corbalán, C. Montoliu, M. Llansola, P. Monfort, S. Erceg et al. 2001. "Aluminium impairs the glutamate-nitric oxide-cGMP pathway in cultured neurons and in rat brain *in vivo*: molecular mechanisms and implications for neuropathology." *Journal of Inorganic Biochemistry* 87:63-9.

Chakraborty, Swarupa, D. K. Kaushik, M. Gupta, and A. Basu. 2010. "Inflammasome signaling at the heart of central nervous system pathology." *Journal of Neuroscience Research* 88:1615-31.

Charlemagne, Daniele, J. Orlowski, P. Oliviero, F. Rannou, C. Sainte Beuve, B. Swynghedauw, and L. K. Lane. 1994. "Alteration of Na,K-ATPase subunit mRNA and protein levels in hypertrophied rat heart." *Journal of Biological Chemistry* 269:1541-7.

Cho, Won Jin, Aleksandar Jeremic, and Bhanu P. Jena. 2005. "Size of supramolecular SNARE complex: membrane-directed self-assembly." *Journal of the American Chemical Society* 127:10156-7.

Colangelo, Vittorio, J. Schurr, M. Ball, R. P. Pelaez, N. G. Bazan, and W. J. Lukiw. 2002. "Gene expression profiling of 12633 genes in Alzheimer hippocampal CA1: Transcription and neurotrophic factor down-regulation and up-regulation." *Journal of Neuroscience Research* 70:462-73.

Cordeiro, João M., S. M. Meireles, M. G. P. Vale, C. R. Oliveira, and P. P. Gonçalves. 2000. "Ca^{2+} regulation of the carrier-mediated γ-aminobutyric acid release from isolated synaptic plasma membrane vesicles." *Neuroscience Research* 38:385-95.

Cordeiro, João M., V. S. Silva, C. R. Oliveira, and P. P. Goncalves. 2003. "Aluminium-induced impairment of Ca^{2+} modulatory action on GABA transport in brain cortex nerve terminals." *Journal of Inorganic Biochemistry* 97:132-42.

Couette, Maryline, M. F. Boisse, P. Maison, P. Brugieres, P. Cesaro, X. Chevalier, R. K. Gherardi et al., 2009. "Long-term persistence of vaccine-derived aluminum hydroxide is associated with chronic cognitive dysfunction." *Journal of Inorganic Biochemistry* 103:1571-8.

Cucarella, Carmen, C. Montoliu, C. Hermenegildo, R. Sáez, L. Manzo, M. D. Miñana, and V. Felipo. 1998. "Chronic exposure to aluminum impairs neuronal glutamatenitric oxide-cyclic GMP pathway." *Journal of Neurochemistry* 70:1609-14.

De Stasio, Gelsomina, M. T. Ciotti, D. Dunham, T. C. Droubay, B. P. Tonner, P. Perfetti, and G. Margaritondo. 1994. "Aluminium in rat cerebellar primary cultures: glial cells and GABAergic neurones." *Neuroreport* 5:1973-6.

Deken, Scott L., M. L. Beckman, L. Boos, and M. W. Quick. 2000. "Transport rates of GABA transporters: regulation by the N-terminal domain and syntaxin 1A." *Nature Neuroscience* 3:998-1003.

Deloncle, Roger, and Olivier Guillard. 1990. "Mechanism of Alzheimer's disease: arguments for a neurotransmitter-aluminium complex implication." *Neurochemical Research* 15: 1239-45.

Dewachter, Ilse, J. Van Dorpe, L. Smeijers, M. Gilis, C. Kuipéri, I. Laenen, N. Caluwaerts et al. 2000. "Aging increased amyloid peptide and caused amyloid plaques in brain of old APP/V717I transgenic mice by a different mechanism than mutant presenilin1." *The Journal of Neuroscience* 20:6452-8.

Diwakar, Latha, and Vijayalakshmi Ravindranath. 2007. "Inhibition of cystathionine-γ-lyase leads to loss of glutathione and aggravation of mitochondrial dysfunction mediated by excitatory amino acid in the CNS." *Neurochemistry International* 50:418-26.

Dunant, Yves. 2006. "Acetylcholine release in rapid synapses." *Journal of Molecular Neuroscience* 30:209-13.

Enokido, Yasushi, E. Suzuki, K. Iwasawa, K. Namekata, H. Okazawa, and H. Kimura. 2005. "Cystathionine β-synthase, a key enzyme for homocysteine metabolism, is preferentially expressed in the radial glia/astrocyte lineage of developing mouse CNS." *The FASEB Journal* 19:1854-6.

Ewing, James F., and Mahin D. Maines. 1992. "*In situ* hybridization and immunohistochemical localization of heme oxygenase-2 mRNA and protein in normal rat brain: differential distribution of isozyme 1 and 2." *Molecular and Cellular Neuroscience* 3:559-70.

Exley, Christopher. 2014. "Why industry propaganda and political interference cannot disguise the inevitable role played by human exposure to aluminum in neurodegenerative diseases, including Alzheimer's disease." *Frontiers in Neurology*. doi:0.3389/ fneur.2014.00212.

Exley, Christopher, and Margaret M. Esiri. 2006. "Severe cerebral congophilic angiopathy coincident with increased brain aluminium in a resident of Camelford, Cornwall, UK." *Journal of Neurology, Neurosurgery and Psychiatry* 77:877-9.

Exley, Christopher, E. House, A. Polwart, and M. M. Esiri. 2012. "Brain burdens of aluminum, iron, and copper and their relationships with amyloid-β pathology in 60 human brains." *Journal of Alzheimer's disease* 31:725-30.

Ferrari, Carlos K. B. 2012. "Effects of xenobiotics on total antioxidant capacity." *Interdisciplinary Toxicology* 5:117-22.

Flašker, Ajda, J. Jorgačevski, A. I. Calejo, M. Kreft, and R. Zorec. 2013. "Vesicle size determines unitary exocytic properties and their sensitivity to sphingosine." *Molecular and Cellular Endocrinology* 376:136-47.

Flendrig, J. Guus, H. Kruis, and H. A. Das, 1976. "Aluminium intoxication: the cause of dialysis dementia?" *Proceedings of the European Dialysis and Transplant Association* 13:355-368.

Fog, Jacob U., H. Khoshbouei, M. Holy, W. A. Owens, C. B. Vaegter, N. Sen, N. Yelyzaveta et al. 2006. "Calmodulin kinase II interacts with the dopamine transporter C terminus to regulate amphetamine-induced reverse transport." *Neuron* 51:417-29.

Fukami, Kazuki, and Atsufumi Kawabata. 2015. "Hydrogen sulfide and neuronal differentiation: Focus on Ca^{2+} channels." *Nitric Oxide* 46:50-4.

Gatta, Valentina, D. Drago, K. Fincati, M. T. Valenti, L. Dalle Carbonare, S. L. Sensi, and P. Zatta. 2011. "Microarray analysis on human neuroblastoma cells exposed to aluminum, β (1-42)-amyloid or the β (1-42)-amyloid aluminum complex." *PloS One* 6: e15965. doi:10.1371/journal.pone.0015965.

Geerlings, Arjan, Beatriz López-Corcuera, and Carmen Aragón. 2000. "Characterization of the interactions between the glycine transporters GLYT1 and GLYT2 and the SNARE protein syntaxin 1A." *FEBS Letters* 470:51-4.

Gherardi, Romain K., and François-Jérôme Authier. 2012. "Macrophagic myofasciitis: characterization and pathophysiology." *Lupus* 21:184-9.

Gherardi, Romain K., M. Coquet, P. Chérin, F.-J. Authier, P. Laforêt, L. Bélec, D. Figarella-Branger et al., 1998. "Macrophagic myofasciitis: an emerging entity." *The Lancet* 352: 347-52.

Gil, Carles, A. Soler-Jover, J. Blasi, and J. Aguilera. 2005. "Synaptic proteins and SNARE complexes are localized in lipid rafts from rat brain synaptosomes." *Biochemical and Biophysical Research Communications* 329:117-24.

Gilbert, M. E., and Timothy J. Shafer. 1996. "*In vitro* exposure to aluminum does not alter long-term potentiation or glutamate release in rat hippocampal slices." *Neurotoxicology and Teratology* 18:175-80.

Gonçalves, Paula P., and Virgília S. Silva. 2007. "Does neurotransmission impairment accompany aluminium neurotoxicity?" *Journal of Inorganic Biochemistry* 101:1291-338.

Gonçalves, Paula P., Sandra M Meireles, and M. Graça P. Vale. 1999. "Regulation of the γ-aminobutyric acid transporter activity by protein phosphatases in synaptic plasma membranes." *Neuroscience Research* 33:41-7.

Guček, Alenka, J. Jorgačevski, U. Górska, B. Rituper, M. Kreft, and R. Zorec. 2015. "Local electrostatic interactions determine the diameter of fusion pores." *Channels* 9:96-101.

Guimarães, Luísa Eça, B. Baker, C. Perricone, and Y. Shoenfeld. 2015. "Vaccines, adjuvants and autoimmunity." *Pharmacological Research* 100:190-209.

Haase, Jana, A. M. Killian, F. Magnani, and C. Williams. 2001. "Regulation of the serotonin transporter by interacting proteins." *Biochemical Society Transactions* 29:722-8.

Hammel, Ilan, and Isaac Meilijson. 2016. "Function suggests nano-structure: Quantitative structural support for SNARE-mediated pore formation." *Neurotoxicity Research Neurotoxicity Research* 29:1-9..

Harrington, Charles R., C. M. Wischik, F. K. McArthur, G. A. Taylor, J. A. Edwardson, and J. M. Candy. 1994. "Alzheimer's disease-like changes in tau protein processing: Association with aluminium accumulation in brains of renal dialysis patients." *Lancet* 343:993-7.

Heinemann, Stefan H., T. Hoshi, M. Westerhausen, and A. Schiller. 2014. "Carbon monoxide–physiology, detection and controlled release." *Chemical Communications* 50:3644-60.

Hermenegildo, Carlos, R. Sáez, C. Minoia, L. Manzo, and V. Felipo. 1999. "Chronic exposure to aluminium impairs the glutamate-nitric oxide-cyclic GMP pathway in the rat *in vivo*." *Neurochemistry International* 34:245-53.

Hofstetter, John R., I. Vincent, O. Bugiani, B. Ghetti, and J. A. Richter. 1987. "Aluminum-induced decreases in choline acetyltransferase, tyrosine hydroxylase, and glutamate decarboxylase in selected regions of rabbit brain." *Neurochemical Pathology* 6:177-93.

Hoozemans, Jeroen J. M., A. J. Rozemuller, E. S. van Haastert, P. Eikelenboom, and W. A. van Gool. 2011. "Neuroinflammation in Alzheimer's disease wanes with age." *Journal of Neuroinflammation* 8:171. doi:10.1186/1742-2094-8-171.

Hoshi, Minako, M. Sato, S. Matsumoto, A. Noguchi, K. Yasutake, N. Yoshida, and K. Sato. 2003. "Spherical aggregates of β-amyloid (amylospheroid) show high neurotoxicity and activate tau protein kinase I/glycogen synthase kinase-3β." *Proceedings of the National Academy of Sciences* 100:6370-5.

Houser, Carolyn R., A. Basil Harris, and James E. Vaughn. 1986. "Time course of the reduction of GABA terminals in a model of focal epilepsy: a glutamic acid decarboxylase immunocytochemical study." *Brain Research* 383:129-45.

Hu, Yunming, Lei Qu, and Thomas Schikorski. 2008. "Mean synaptic vesicle size varies among individual excitatory hippocampal synapses." *Synapse* 62: 953-7.

Imuta, Naohiko, O. Hori, Y. Kitao, Y. Tabata, T. Yoshimoto, T. Matsuyama, and S. Ogawa. 2007. "Hypoxia-mediated induction of heme oxygenase type I and carbon monoxide release from astrocytes protects nearby cerebral neurons from hypoxia-mediated apoptosis." *Antioxidants and Redox Signaling* 9:543-52.

Ishii, Isao, N. Akahoshi, X. Yu, Y. Kobayashi, K. Namekata, G. Komaki, and H. Kimura. 2004. "Murine cystathionine gamma-lyase: complete cDNA and genomic sequences, promoter activity, tissue distribution and developmental expression." *Biochemical Journal* 381:113-23.

Israël, Maurice, B. Lesbats, N. Morel, R. Manaranche, and G. L. G. la Salle. 1988. "Is the acetylcholine releasing protein mediatophore present in rat brain?" *FEBS Letters* 233:421-6.

Jorgačevski, Jernej, M. Fošnarič, N. Vardjan, M. Stenovec, M. Potokar, M. Kreft, V. Kralj-Iglič et al. 2010. "Fusion pore stability of peptidergic vesicles." *Molecular Membrane Biology* 27:65-80.

Jorgačevski, Jernej, M. Kreft, N. Vardjan, and R. Zorec. 2012. "Fusion pore regulation in peptidergic vesicles." *Cell Calcium* 52:270-6.

Julka, Deepinder, Rajat Sandhir, and Kiran Dip Gill. 1995. "Altered cholinergic metabolism in rat CNS following aluminum exposure: implications on learning performance." *Journal of Neurochemistry* 65:2157-64.

Kabaso, Doron, A. I. Calejo, J. Jorgačevski, M. Kreft, R. Zorec, and A. Iglič. 2012. "Fusion pore diameter regulation by cations modulating local membrane anisotropy." *The Scientific World Journal*. doi:10.1100/2012/983138.

Kaur, Amarpreet, and Kiran Dip Gill. 2005. "Disruption of neuronal calcium homeostasis after chronic aluminium toxicity in rats." *Basic and Clinical Pharmacology and Toxicology* 96:118-22.

Kawahara, Masahiro, K. Muramoto, K. Kobayashi, H. Mori, and Y. Kuroda. 1994. "Aluminum promotes the aggregation of Alzheimer's amyloid β-protein *in vitro*." *Biochemical and Biophysical Research Communications* 198:531-5.

Kawahara, Masahiro, and Midori Kato-Negishi. 2011. "Link between aluminum and the pathogenesis of Alzheimer's disease: the integration of the aluminum and amyloid cascade hypotheses." *International Journal of Alzheimer's Disease* 2011. doi:10.4061/2011/276393.

Kida, Kotaro, M. Yamada, K. Tokuda, E. Marutani, M. Kakinohana, M. Kaneki, and F. Ichinose. 2011. "Inhaled hydrogen sulfide prevents neurodegeneration and movement disorder in a mouse model of Parkinson's disease." *Antioxidants and Redox Signaling* 15:343-52.

Kim, Kisok. 2003. "Perinatal exposure to aluminum alters neuronal nitric oxide synthase expression in the frontal cortex of rat offspring." *Brain Research Bulletin* 61:437-41.

Kim, Sunyoun, Jungmin Nam, and Kisok Kim. 2007. "Aluminum exposure decreases dopamine D1 and D2 receptor expression in mouse brain." *Human and Experimental Toxicology* 26:741-6.

Kimura, Hideo. 2013. "Physiological role of hydrogen sulfide and polysulfide in the central nervous system." *Neurochemistry International* 63:492-7.

Kimura, Hideo. 2014. "The physiological role of hydrogen sulfide and beyond." *Nitric Oxide* 41:4-10.

Kimura, Yuka, Yu-Ichi Goto, and Hideo Kimura. 2010. "Hydrogen sulfide increases glutathione production and suppresses oxidative stress in mitochondria." *Antioxidants and Redox Signaling* 12:1-13.

Klyachko, Vitaly A., and Meyer B. Jackson. 2002. "Capacitance steps and fusion pores of small and large-dense-core vesicles in nerve terminals." *Nature* 418:89-92.

Kodavanti, Prasada, S. Rao, W. R. Mundy, H. A. Tilson, and G. J. Harry. 1993. "Effects of selected neuroactive chemicals on calcium transporting systems in rat cerebellum and on survival of cerebellar granule cells." *Toxicological Sciences* 21:308-16.

Konturek, Stanislaw J., T. Brzozowski, J. Majka, A. Szlachcic, C. Nauert, and B. Slomiany. 1992. "Nitric oxide in gastroprotection by aluminium-containing antacids." *European Journal of Pharmacology* 229:155-62.

Kumar, Shantanu. 2002. "Aluminium-induced changes in the rat brain serotonin system." *Food and Chemical Toxicology* 40:1875-80.

Kumar, Vijay, and Kiran Dip Gill. 2014. "Oxidative stress and mitochondrial dysfunction in aluminium neurotoxicity and its amelioration: a review." *Neurotoxicology* 41:154-66.

Laabbar, Wafaa, A. Elgot, N. Kissani, and H. Gamrani. 2014. "Chronic aluminum intoxication in rat induced both serotonin changes in the dorsal raphe nucleus and alteration of glycoprotein secretion in the subcommissural organ: Immunohistochemical study." *Neuroscience Letters* 577:72-6.

Lai, James C. K., Louis Lim, and Alan N. Davison. 1981. "Differences in the inhibitory effect of Cd^{2+}, Mn^{2+}, and Al^{3+} on the uptake of dopamine by synaptosomes from forebrain and from striatum of the rat." *Biochemical Pharmacology* 30:3123-5.

Lai, James C. K., Louis Lim, and Alan N. Davison. 1982. "Effects of Cd^{2+}, Mn^{2+}, and Al^{3+} on rat brain synaptosomal uptake of noradrenaline and serotonin." *Journal of Inorganic Biochemistry* 17:215-25.

Lang, Thorsten, D. Bruns, D. Wenzel, D. Riedel, P. Holroyd, C. Thiele, and R. Jahn. 2001. "SNAREs are concentrated in cholesterol-dependent clusters that define docking and fusion sites for exocytosis." *The EMBO Journal* 20:2202-13.

Levi, Giulio, and Maurizio Raiteri. 1993. "Carrier-mediated release of neurotransmitters." *Trends in Neurosciences* 16:415-9.

Li, Wei, K. K. Ma, W. Sun, and H. K. Paudel. 1998. "Phosphorylation sensitizes microtubule-associated protein τ to Al^{3+}-induced aggregation." *Neurochemical Research* 23:1467-76.

Li, Yinxia, S. Yu, Q. Wu, M. Tang, and D. Wang. 2013. "Transmissions of serotonin, dopamine, and glutamate are required for the formation of neurotoxicity from Al_2O_3-NPs in nematode *Caenorhabditis elegans.*" *Nanotoxicology* 7:1004-13.

Llansola, Marta, M. D. Miñana, C. Montoliu, R. Saez, R. Corbalán, L. Manzo, and V. Felipo. 1999. "Prenatal exposure to aluminum reduces expression of neuronal nitric oxide synthase and of soluble guanylate cyclase and impairs glutamatergic neurotransmission in rat cerebellum." *Journal of Neurochemistry* 73:712-8.

Lukiw, Walter J., Maire E. Percy, and Theo P. Kruck. 2005. "Nanomolar aluminum induces pro-inflammatory and pro-apoptotic gene expression in human brain cells in primary culture." *Journal of Inorganic Biochemistry* 99:1895-98.

Middelhaufe, Sabine, M. Leipelt, L. R. Levin, J. Buck, and C. Steegborn. 2012. "Identification of a haem domain in human soluble adenylate cyclase." *Bioscience Reports* 32:491-9.

Motterlini, Roberto, R. Foresti, M. Intaglietta, and R. M. Winslow. 1996. "NO-mediated activation of heme oxygenase: Endogenous cytoprotection against oxidative stress to endothelium." *American Journal of Physiology-Heart and Circulatory Physiology* 39: H107-H114.

Mundy, William R., Theresa M. Freudenrich, and Prasad R. S. Kodavanti. 1997. "Aluminum potentiates glutamate-induced calcium accumulation and iron-induced oxygen free radical formation in primary neuronal cultures." *Molecular and Chemical Neuropathology* 32:41-57.

Nagahara, Noriyuki. 2013. "Regulation of mercaptopyruvate sulfurtransferase activity via intrasubunit and intersubunit redox-sensing switches." *Antioxidants and Redox Signaling* 19:1792-802.

Nayak, Prasunpriya, and Ajay K. Chatterjee. 2001. "Effects of aluminium exposure on brain glutamate and GABA systems: an experimental study in rats." *Food and Chemical Toxicology* 39:1285-9.

Nday, Christiane M., B. D. Drever, T. Salifoglou, and B. Platt. 2010. "Aluminium interferes with hippocampal calcium signaling in a species-specific manner." *Journal of Inorganic Biochemistry* 104:919-27.

Neill, David, A. Leake, D. Hughes, A. B. Keith, G. A. Taylor, D. Allsop, B. K. Rima et al. 1996. "Effect of aluminum on expression and processing of amyloid precursor protein." *Journal of Neuroscience Research* 46:395-403.

Nelson, Lucy, and Naji Tabet. 2015. "Slowing the progression of Alzheimer's disease; what works?" *Ageing Research Reviews* 23:193-209.

Nichols, Robert A., Gregory R. Suplick, and Jacquelyn M. Brown. 1994. "Calcineurin-mediated protein dephosphorylation in brain nerve terminals regulates the release of glutamate." *Journal of Biological Chemistry* 269:23817-23.

Ohnishi, Takayuki, M. Yanazawa, T. Sasahara, Y. Kitamura, H. Hiroaki, Y. Fukazawa, I. Kiif et al. 2015. "Na, K-ATPase α3 is a death target of Alzheimer patient amyloid-β assembly." *Proceedings of the National Academy of Sciences* 112:E4465-74.

Passeri, Elodie, C. Villa, M. Couette, E. Itti, P. Brugieres, P. Cesaro, R. K. Gherardi et al. 2011. "Long-term follow-up of cognitive dysfunction in patients with aluminum hydroxideinducedmacrophagicmyofasciitis (MMF)." *Journal of Inorganic Biochemistry* 105:1457-63.

Pearson, Hugh A., and Chris Peers. 2006. "Physiological roles for amyloid β peptides." *The Journal of Physiology* 575:5-10.

Percy, Maire E., T. P. Kruck, A. I. Pogue, and W. J. Lukiw. 2011. "Towards the prevention of potential aluminum toxic effects and an effective treatment for Alzheimer's disease." *Journal of Inorganic Biochemistry* 105:1505-12.

Pineau, Alain, B. Fauconneau, A. P. Sappino, R. Deloncle, and O. Guillard. 2014. "If exposure to aluminium in antiperspirants presents health risks, its content should be reduced." *Journal of Trace Elements in Medicine and Biology* 28:147-50.

Pivnick, Enikö K., N. C. Kerr, R. A. Kaufman, D. P. Jones, and R.W. Chesney. 1995. "Rickets secondary to phosphate depletion: a sequela of antacid use in infancy." *Clinical Pediatrics* 34:73-8.

Platt, Bettina, D. O. Carpenter, D. Büsselberg, K. G. Reymann, and G. Riedel. 1995. "Aluminum impairs hippocampal long-term potentiation in rats *in vitro* and *in vivo*." *Experimental Neurology* 134:73-86.

Platt, Bettina, G. Fiddler, G. Riedel, and Z. Henderson. 2001. "Aluminium toxicity in the rat brain: histochemical and immunocytochemical evidence." *Brain Research Bulletin* 55: 257-67.

Prakash, Dharmalingam, Kulasekaran Gopinath, and Ganapasam Sudhandiran. 2013. "Fisetin enhances behavioral performances and attenuates reactive gliosis and inflammation during aluminum chloride-induced neurotoxicity." *Neuromolecular Medicine* 15: 192-208.

Praticò, Domenico, K. Uryu, S. Sung, S. Tang, J. Q. Trojanowski, and V. M. Lee. 2002. "Aluminum modulates brain amyloidosis through oxidative stress in APP transgenic mice." *The FASEB Journal* 16:1138-40.

Provan, Spencer D., and Robert A. Yokel. 1992. "Aluminium inhibits glutamate release from transverse rat hippocampal slices: role of G proteins, Ca channels and protein kinase C." *Neurotoxicology* 13:413-20.

Raiteri, Luca, and Maurizio Raiteri. 2015. "Multiple functions of neuronal plasma membrane neurotransmitter transporters." *Progress in Neurobiology* 134:1-16.

Ramamoorthy, Sammanda, Toni S. Shippenberg, and Lankupalle D. Jayanthi. 2011. "Regulation of monoamine transporters: Role of transporter phosphorylation." *Pharmacology and Therapeutics* 129:220-38.

Reese, Lindsay C., and Giulio Taglialatela. 2011. "A role for calcineurin in Alzheimer's disease." *Current Neuropharmacology* 9:685-92.

Reusche, Erich, P. Pilz, G. Oberascher, B. Lindner, R. Egensperger, K. Gloeckner, E. Trinka, and B. Iglseder. 2001. "Subacute fatal aluminum encephalopathy after reconstructive otoneurosurgery: a case report." *Human Pathology* 32:1136-40.

Rituper, Boštjan, A. Flašker, A. Guček, H. H. Chowdhury, and R. Zorec. 2012. "Cholesterol and regulated exocytosis: a requirement for unitary exocytotic events." *Cell Calcium* 52:250-8.

Rodella, Luigi F., F. Ricci, E. Borsani, R. Rezzani, A. Stacchiotti, C. Mariani, and R. Bianchi. 2006. "Exposure to aluminium changes the NADPH-diaphorase/NPY pattern in the rat cerebral cortex." *Archives of Histology and Cytology* 69:13-21.

Rondeau, Virginie, D. Commenges, H. Jacqmin-Gadda, and J. F. Dartigues. 2000. "Relation between aluminum concentrations in drinking water and Alzheimer's disease: an 8-year follow-up study." *American Journal of Epidemiology* 152:59-66.

Rossi, David J., Takeo Oshima, and David Attwell. 2000. "Glutamate release in severe brain ischaemia is mainly by reversed uptake." *Nature* 403:316-21.

Ryan, Aisling M., N. Bermingham, H. J. Harrington, and C. Keohane. 2006. "Atypical presentation of macrophagic myofasciitis 10 years post vaccination." *Neuromuscular Disorders*, 16:867-9.

Satoh, Eiko, I. Yasuda, T. Yamada, Y. Suzuki, and T. Ohyashiki. 2007. "Involvement of NO generation in aluminum-induced cell death." *Biological and Pharmaceutical Bulletin* 30:1390-4.

Schipper, Hyman M., H. Chertkow, K. Mehindate, D. Frankel, C. Melmed, and H. Bergman. 2000. "Evaluation of heme oxygenase-1 as a systemic biological marker of sporadic AD." *Neurology* 54:1297-304.

Segatto, Marco, L. Leboffe, L. Trapani, and V. Pallottini. 2014. "Cholesterol homeostasis failure in the brain: implications for synaptic dysfunction and cognitive decline." *Current Medicinal Chemistry* 21:2788-802.

Selkoe, Dennis J. 2002. "Alzheimer's disease is a synaptic failure." *Science* 298:789-91.

Serluca, Fabrizio C., A. Sidow, J. D. Mably, and M. C. Fishman. 2001. "Partitioning of tissue expression accompanies multiple duplications of the Na^+/K^+ ATPase α subunit gene." *Genome Research* 11:1625-31.

Seubert, Anja, E. Monaci, M. Pizza, D. T. O'Hagan, and A. Wack. 2008. "The adjuvants aluminium hydroxide and MF59 induce monocyte and granulocyte chemoattractants and enhance monocyte differentiation toward dendritic cells." *The Journal of Immunology* 180:5402-12.

Shaw, Christopher A., S. Seneff, S. D. Kette, L. Tomljenovic, J. W. Jr. Oller, and R. M. Davidson. 2014. "Aluminum-induced entropy in biological systems: implications for neurological disease." *Journal of Toxicology* 2014. http://dx.doi.org/10.1155/2014/491 316.

Shibuya, Norihiro, M. Tanaka, M. Yoshida, Y. Ogasawara, T. Togawa, K. Ishii, and H. Kimura. 2009. "3-Mercaptopyruvate sulfurtransferase produces hydrogen sulfide and bound sulfane sulfur in the brain." *Antioxidants and Redox Signaling* 11:703-14.

Shuchang, He, N. Qiao, N. Piye, H. Mingwei, S. Xiaoshu, S. Feng, W. Sheng, and M. Oplerf. 2008. "Protective effects of gastrodia elata on aluminium-chloride-induced learning impairments and alterations of amino acid neurotransmitter release in adult rats." *Restorative Neurology and Neuroscience* 26:467-73.

Silva, Virgília S., and Paula P. Gonçalves. 2003. "The inhibitory effect of aluminium on the (Na^+/K^+)ATPase activity of rat brain cortex synaptosomes." *Journal of Inorganic Biochemistry* 97:143-50.

Silva, Virgília S., and Paula P. Gonçalves. 2015. "Effect of lysine acetylsalicylate on aluminium accumulation and (Na^+/K^+)ATPase activity in rat brain cortex synaptosomes after aluminium ingestion." *Toxicology Letters* 232:167-74.

Silva, Virgília S., J. M. Cordeiro, M. J Matos, C. R. Oliveira, P. P. Gonçalves. 2002. "Aluminum accumulation and membrane fluidity alteration in synaptosomes isolated from rat brain cortex following aluminum ingestion: effect of cholesterol." *Neuroscience Research* 44:181-93.

Silva, Virgília S., A. I. Duarte, A. C. Rego, C. R. Oliveira, and P. P. Gonçalves. 2005. "Effect of chronic exposure to aluminium on isoform expression and activity of rat (Na^+/K^+) ATPase." *Toxicological Sciences* 88:485-94.

Silva, Virgília S., M. A. Nunes, J. M. Cordeiro, A. I. Calejo, S. Santos, P. Neves, A. Sykes et al. 2007. "Comparative effects of aluminum and ouabain on synaptosomal choline uptake, acetylcholine release and (Na^+/K^+)ATPase." *Toxicology* 236:158-77.

Silva, Virgília S., Liliana Oliveira, and Paula P. Gonçalves. 2013. "Alteration of aluminium inhibition of synaptosomal (Na^+/K^+) ATPase by colestipol administration." *Journal of Inorganic Biochemistry* 128:208-14.

Sinczuk-Walczak, Halina, M. Szymczak, G. Raźniewska, W. Matczak, and W. Szymczak. 2003. "Effects of occupational exposure to aluminum on nervous system: clinical and electroencephalographic findings." *International Journal of Occupational Medicine and Environmental Health* 16:301-10.

Song, Jing, Y. Liu, H. F. Zhang, Q. L. Zhang, and Q. Niu. 2014. "Effects of exposure to aluminum on long-term potentiation and AMPA receptor subunits in rats in *vivo*." *Biomedical and Environmental Sciences* 27:77-84.

Stamler, Jonathan S., D. I. Simon, J. A. Osborne, M. E. Mullins, O. Jaraki, T. Michel, D. J. Singel, and J. Loscalzo. 1992. "S-nitrosylation of proteins with nitric oxide: synthesis and characterization of biologically active compounds." *Proceedings of the National Academy of Sciences* 89:444-8.

Steiner, Jennifer A., Ana Marin D. Carneiro, and Randy D. Blakely. 2008. "Going with the flow: trafficking-dependent and-independent regulation of serotonin transport." *Traffic* 9:1393-402.

Stenovec, Matjaž, S. Trkov, E. Lasič, S. Terzieva, M. Kreft, J. J. Rodríguez Arellano, P. Vladimir Parpura et al. 2016. "Expression of familial Alzheimer disease presenilin 1 gene attenuates vesicle traffic and reduces peptide secretion in cultured astrocytes devoid of pathologic tissue environment." *Glia* 64:317-29.

Stevanović, Ivana D., M. D. Jovanović, M. Čolić, A. Jelenković, D. Bokonjić, M. Ninković, and I. Stojanović. 2011. "N-nitro-L-arginine methyl ester influence on aluminium toxicity in the brain." *Folia Neuropathologica* 49:219-29.

Stevanović, Ivana D., M. D. Jovanović, A. Jelenković, M. Čolić, I. Stojanović, and M. Ninković. 2009a. "Effects of L-NAME, a non-specific nitric oxide synthase inhibitor, on $AlCl_3$-induced toxicity in the rat forebrain cortex." *Journal of Veterinary Science* 10:15-22.

Stevanović, Ivana D., M. D. Jovanović, A. Jelenković, M. Ninković, M. Đukić, I. Stojanović, and M. Čolić. 2009b. "The effect of inhibition of nitric oxide synthase on aluminium-induced toxicity in the rat brain." *General Physiology and Biophysics* 28:235-42.

Stone, James R., and Michael A. Marletta. 1994. "Soluble guanylate cyclase from bovine lung: activation with nitric oxide and carbon monoxide and spectral characterization of the ferrous and ferric states." *Biochemistry* 33:5636-40.

Struys-Ponsar, C., Olivier Guillard, and Philippe van den Bosch de Aguilar. 2000. "Effects of aluminum exposure on glutamate metabolism: a possible explanation for its toxicity." *Experimental Neurology* 163:157-64.

Sturman, John A., Henryk M. Wisniewski, and Judy W. Shek. 1983. "High affinity uptake of GABA and glycine by rabbits with aluminum-induced neurofibrillary changes." *Neurochemical Research* 8:1097-109.

Sumathi, Thangarajan, C. Shobana, M. Thangarajeswari, and R. Usha. 2014. "Protective effect of L-Theanine against aluminium induced neurotoxicity in cerebral cortex, hippocampus and cerebellum of rat brain-histopathological, and biochemical approach." *Drug and Chemical Toxicology* 38:22-31.

Sung, Uhna, S. Apparsundaram, A. Galli, K. M. Kahlig, V. Savchenko, S. Schroeter, M.W., Quick, and R. D. Blakely. 2003. "A regulated interaction of syntaxin 1A with the antidepressant-sensitive norepinephrine transporter establishes catecholamine clearance capacity." *The Journal of Neuroscience* 23:1697-709.

Sweeney, Vincent P., T. L. Perry, J. D. Price, C. E. Reeve, W. J. Godolphin, and S. J. Kish. 1985. "Brain γ-aminobutyric acid deficiency in dialysis encephalopathy." *Neurology* 35:180-180.

Szutowicz, Andrzej. 2001. "Aluminum, NO, and nerve growth factor neurotoxicity in cholinergic neurons." *Journal of Neuroscience Research* 66:1009-18.

Szutowicz, Andrzej, M. Tomaszewicz, H. Bielarczyk, and A. Jankowska. 1998. "Putative significance of shifts in acetyl-CoA compartmentalization in nerve terminals for disturbances of cholinergic transmission in brain." *Developmental Neuroscience* 20: 485-92.

Trombley, Paul Q. 1998. "Selective modulation of $GABA_A$ receptors by aluminum." *Journal of Neurophysiology* 80:755-61.

Tsunoda, Masashi, and Raghubir P. Sharma. 1999. "Altered dopamine turnover in murine hypothalamus after low-dose continuous oral administration of aluminum." *Journal of Trace Elements in Medicine and Biology* 13:224-31.

Vardjan, Nina, Jernej Jorgačevski, and Robert Zorec. 2013. "Fusion pores, SNAREs, and exocytosis." *The Neuroscientist* 19:160-74.

Vardjan, Nina, Alexei Verkhratsky, and Robert Zorec. 2015. "Pathologic potential of astrocytic vesicle traffic: new targets to treat neurologic diseases?" *Cell Transplantation* 24:599-612.

Verstraeten, Sandra V., Lucila Aimo, and Patricia I. Oteiza. 2008. "Aluminium and lead: molecular mechanisms of brain toxicity." *Archives of Toxicology* 82:789-802.

Walton, Judie R. 2012. "Aluminum disruption of calcium homeostasis and signal transduction resembles change that occurs in aging and Alzheimer's disease." *Journal of Alzheimer's Disease* 29:255-73.

Walton, Judie R. 2014. "Chronic aluminum intake causes Alzheimer's disease: Applying Sir Austin Bradford Hill's causality criteria." *Journal of Alzheimer's Disease* 40:765-838.

Wang, Linping, J. Hu, Y. Zhao, X. Lu, Q. Zhang, and Q. Niu. 2014. "Effects of aluminium on β-amyloid (1-42) and secretases (APP-cleaving enzymes) in rat brain." *Neurochemical Research* 39:1338-45.

Wen, Haitao, D. Gris, Y. Lei, S. Jha, L. Zhang, M. T.-H. Huang, W. J. Brickey, and J. P-Y. Ting. 2011. "Fatty acid-induced NLRP3-ASC inflammasome activation interferes with insulin signaling." *Nature Immunology* 12:408-15.

Willhite, Calvin C., N. A. Karyakina, R. A. Yokel, N. Yenugadhati, T. M. Wisniewski, I. M. Arnold, F. Momolifgh and D. Krewskiabg. 2014. "Systematic review of potential health risks posed by pharmaceutical, occupational and consumer exposures to metallic and nanoscale aluminum, aluminum oxides, aluminum hydroxide and its soluble salts." *Critical Reviews in Toxicology* 44:1-80.

Wong, Patrick C. L., J. C. K. Lai, L. Lim, and A. N. Davidson. 1981. "Selective inhibition of L-glutamate and gammaaminobutyrate transport in nerve ending particles by aluminium, manganese and cadmium chloride." *Journal of Inorganic Biochemistry* 14: 253-60.

Wu, Lingyun, and Rui Wang. 2005. "Carbon monoxide: endogenous production, physiological functions, and pharmacological applications." *Pharmacological Reviews* 57:585-630.

Xuan, Aiguo, D. Long, J. Li, W. Ji, M. Zhang, L. Hong, and J. Liu. 2012. "Hydrogen sulfide attenuates spatial memory impairment and hippocampal neuroinflammation in beta-amyloid rat model of Alzheimer's disease." *Journal of Neuroinflammation* 9:202. doi: 10. 1186/1742-2094-9-202.

Yang, Ellen Y., Shirley X. Guo-Ross, and Stephen C. Bondy. 1999. "The stabilization of ferrous iron by a toxic β-amyloid fragment and by an aluminum salt." *Brain Research* 839:221-6.

Yang, Guangdong, L. Wu, B. Jiang, W. Yang, J. Qi, K. Cao, Q. Meng et al. 2008. "H_2S as a physiologic vasorelaxant: hypertension in mice with deletion of cystathionine γ-lyase." *Science* 322:587-90.

Yang, Yoosoo, J. Kim, H. Y. Kim, N. Ryoo, S. Lee, Y. Kim, H. Rhim, and Y. K. Shin. 2015. "Amyloid-β oligomers may impair SNARE-mediated exocytosis by direct binding to syntaxin 1a." *Cell Reports* 12:1244-51.

Yokel, Robert A. 2006. "Blood-brain barrier flux of aluminum, manganese, iron and other metals suspected to contribute to metal-induced neurodegeneration." *Journal of Alzheimer's Disease* 10:223-53.

Yokel, Robert A., S. D. Provan, J. J. Meyer, and S. R. Campbell. 1987. "Aluminum intoxication and the victim of Alzheimer's disease: similarities and differences." *Neurotoxicology* 9:429-42.

Yu, Jin-Tai, Raymond C. C. Chang, and Lan Tan. 2009. "Calcium dysregulation in Alzheimer's disease: from mechanisms to therapeutic opportunities." *Progress in Neurobiology* 89:240-55.

Zaky, Amira, B. Mohammad, M. Moftah, K. M. Kandeel, and A. R. Bassiouny. 2013. "Apurinic/apyrimidinic endonuclease 1 is a key modulator of aluminum-induced neuroinflammation." *BMC Neuroscience* 14:26. doi:10.1186/1471-2202-14-26.

Zatta, Paolo, P. Zambenedetti, E. Reusche, F. Stellmacher, A. Cester, P. Albanese, G. Meneghel, and M. Nordio. 2004. "A fatal case of aluminium encephalopathy in a patient with severe chronic renal failure not on dialysis." *Nephrology Dialysis Transplantation.* 19:2929-31.

Zhang, Fuping, L. Yang S. Bi J. Liu F. Liu X. Wang X. Yang et al. 2001. "Neurotransmitter dopamine applied in electrochemical determination of aluminum in drinking waters and biological samples." *Journal of Inorganic Biochemistry* 87:105-13.

Zhang, Lifeng, C. Jin, X. Lu, J. Yang, S.Wu, Q. Liu, R. Chen et al. 2014. "Aluminium chloride impairs long-term memory and downregulates cAMP-PKA-CREB signalling in rats." *Toxicology* 323:95-108.

Zhang, Zhen, Y. Wu, Z. Wang, F. M. Dunning, J. Rehfuss, D. Ramanan, E. R. Chapman, and M. B. Jackson. 2011. "Release mode of large and small dense-core vesicles specified by different synaptotagmin isoforms in PC12 cells." *Molecular Biology of the Cell* 22: 2324-36.

Zhao, Minglei, S. Wu, Q. Zhou, S. Vivona, D. J. Cipriano, Y. Cheng, and A. T. Brunger. 2015. "Mechanistic insights into the recycling machine of the SNARE complex." *Nature* 518:61-7.

Zhu, Yanzhu, Y. Li, L. Miao, Y. Wang, Y. Liu, X. Yan, X. Cui, and H. Li. 2014. "Immunotoxicity of aluminum." *Chemosphere* 104:1-6.

Zielke, H. Ronald, M. J. Jackson, J. T. Tildon, and S. R. Max. 1993. "A glutamatergic mechanism for aluminum toxicity in astrocytes." *Molecular and Chemical Neuropathology* 19:219-33.

In: Aluminum Neurotoxicity
Editor: Ankica Jelenković
ISBN: 978-1-63484-739-1

Chapter 3

THE RISK OF ALUMINIUM NEUROTOXICITY FOR YOUNG ANIMALS AND HUMANS DUE TO MULTIPLE EXPOSURE OPPORTUNITIES, PARTICULARLY PERINATAL

Mohammad Ahmad[1,*], Gasem M. Abu-Taweel[2] and Jamaan S. Ajarem[3]
[1]Department of Medical Surgical Nursing, College of Nursing, King Saud University, Riyadh, Saudi Arabia
[2]Department of Biology, College of Education, Dammam University, Dammam, Saudi Arabia
[3]Department of Zoology, College of Science, King Saud University, Riyadh, Saudi Arabia

ABSTRACT

Aluminium (Al) is a well-known neurotoxicant in adults as well as in developing stages of animals and humans. Al is reported in the etiology of several neurodegenerative disorders like Parkinson's disease, dementia, and Alzheimer's disease. Besides multiple exposure opportunities for early exposure of the fetus to Al, the route or mode of exposure is also critical for the developing offspring. Al-induced behavioral alterations as well as cognitive deficits and rodent brain neurotransmitters level, have been widely reported in literature but exact mechanism in the offspring of perinatally Al exposed dams is not yet understood properly and needs more attention. In the present study, the perinatal oral exposure of the dams to 300 and 600 mg/kg Al (aluminium chloride) resulted in a significant and dose-dependent reduction in postnatal body weight gain, delays in opening of the eyes and appearance in the body hair fuzz, and deficits in the sensory motor reflexes of the mice pups during weaning period (from the day of birth to postnatal day 21). At adolescent and adult ages of the male offspring, a significant and dose-dependent deficit was also observed in their learning capability (at postnatal day-

* Corresponding author: Tel: +966 505195887; Fax: +966 11 4693625. E-mail address: mbadshah@ksu.edu.sa (M. Ahmad).

PD25), and cognitive behavior (at PD30-36). Furthermore, a significant and dose-dependent disturbance in the levels of neurotransmitters like dopamine (DA) and serotonin (5-HT); non enzymatic oxidative stress (OS) indices like thiobarbituric acid-ractive substances (TBARS) and total reduced glutathione (GSH); and enzymatic OS indices like glutathione S-transferase (GST), catalase (CAT), and superoxide dismutase (SOD) were observed in the forebrain region of the offspring at post-natal day (PD)7, PD14, PD21, PD30, and PD36. Thus, perinatal Al exposure can affect the in utero developing fetus, raising the concerns that during a critical prenatal and postnatal period of brain development, Al exposure has potential neurotoxic hazards and might modify the properties of the dopaminergic system and inflicts OS that changes the threshold of that system or other related systems resulting into a longer lasting cognitive dysfunction. A reduced use of Al during pregnancy is of crucial importance in preventing Al-induced neurotoxicity in the offspring.

Keywords: perinatal, aluminium, mice offspring, neurotoxicity, cognitive dysfunction, brain, neurotransmitters, oxidative stress

INTRODUCTION

Aluminum (Al) neurotoxicity is well known in adults (Golub and Germann, 2001) as well as in developing stages of altricial animals like rats and mice (Domingo, 1995; Golub and Domingo, 1996). Early exposure to Al in life may lead to increased functional impairments and potential developmental neurotoxicities (Reuhl, 1991). Al has been linked with the etiology of several neurodegenerative disorders like Parkinson's disease (Yasui et al., 1992; Sanchez-Iglesias et al., 2009), dementia (Forbes et al., 1995), and Alzheimer's disease (Kawahara, 2005; Mahdi et al., 2006), however, linkages to the findings outlined above do not convincingly demonstrate a causal relationship between Al and neurodegenerative diseases (Massey and Taylor, 1989; Exley, 2001). It is argued that not the increased exposure to Al alone but the elevated level of Al may be the factor in initiating and aggravating the pathogenesis of neurodegenerative diseases (Sethi et al., 2008), but still it is not generally accepted that Al is a contributory factor in the etiology of such neurodegenerative diseases because the precise mechanism of such disease pathogenesis remains unknown, and still this issue is controversial (Kawahara and Kato-Negishi, 2011).

Although multiple exposure opportunities lie for early exposure of the fetus to Al, which may include prenatal, postnatal and perinatal exposures, perinatal period is critically most important from the brain development point of view. Perinatal involves prenatal as well as postnatal periods and a major portion of brain cells (70%) of rodents are known to form postnatally after birth (Patel, 1983). The brain is usually susceptible to disruption by drugs during its development, with drugs potentially affecting a variety of developmental processes, including neuronal migration and differentiation, glia proliferation, along with normal ontogenetic changes in cell adhesion, neural communications, energy utilization, and apoptosis (Goodlett and Horn, 2001). Different portions of the brain develop at different time and hence differ in the timing of their vulnerability to drug-induced disruptions. Thus, consequences of developmental exposures to drugs are to some extent dependent on the timings of the exposure. Perinatal period in rodents is approximately equivalent to the first and second trimesters of human pregnancy, with the first 10 days or so of postnatal life in rats, approximating the third human trimester (Spear and File, 1996). It is for this reason that we consider here to discuss the deleterious neurotoxic effects of Al in mice offspring that are exposed to Al during perinatal period.

Besides exposure opportunities, the routes or modes of Al exposure also affect the developmental neurotoxicity and many of such perinatal studies have been carried out by exposing the animals through Al injection during early pregnancy; in spite of the fact that oral administration appears to be more readily common in human populations. Furthermore, orally ingested Al consumed through foods, water, freshly-prepared natural sources (mainly fresh fruit, vegetables and meat), and as additive in commercially-processed foods and beverages (Greger, 1993; Walton, 2007), currently comprises the main form of Al exposure for the general public (WHO, 2006). Also, Al gets an easy access to our body through use of cooking utensils, deodorants, antacids, etc. (Yokel, 2000).

Although long-term effects of developmental exposure (perinatal) has not been studied much in this regard in human populations, a recent study detected a delay in neuromaturation in preterm infants exposed to high levels of Al through parenteral fluids (Bishop et al., 1997). Behavioral effects of developmental (perinatal) exposure to Al have been examined to some extent using laboratory rodents (Colomina et al., 1999; Gonda and Lehotzky, 1996; Gonda et al., 1997; Santucci et al., 1994; Abu-Taweel et al., 2012). Maternal stress during pregnancy can have serious negative effects on the learning abilities of the offspring (Nishio et al., 2001; Sternberg and Ridgway, 2003). Exposure to Al during pregnancy along with maternal stress can further have serious interactions and result into learning defects in the offspring (Colomina et al., 2005; Roig et al., 2006). Al has been demonstrated to be capable not only of crossing the blood brain barrier (BBB), but also of increasing its permeability (Zatta et al., 2002). It is well established that Al is toxic to the growth and development of fetuses and suckling in experimental animals (Domingo, 1995; Sharma and Mishra, 2006). Moreover, it has been shown that gestational and lactational exposure to doses of Al that do not produce maternal toxicity can result in persistent neurobehavioral deficits in the offspring of some mammals (Colomina et al., 2005; Abu-Taweel et al., 2012). These Al doses may not correspond to the level of exposure in humans, but its toxic effects reported in animal studies have a significant relevance to Al exposure in humans (Abu-Taweel et al., 2012). Although Al induced behavioral alterations as well as cognitive deficits have been reported in literature, the effects in the offspring of perinatally exposed Al dams are not yet understood properly and need more attention in order to gain information on the longer lasting effects of Al transferred from dams to offspring. Furthermore, Al induced changes in the rodent brain neurotransmitters studies are scanty (Kumar, 2002). Al administration has been reported to induce oxidative stress (OS) by inflicting damage to membrane lipid, proteins and antioxidative enzyme defense system (Jyoti et al., 2007). Furthermore, Al has been associated with the OS in brain of adult rodents which in turn has been linked with neurodegenerative diseases and cognitive dysfunctions (Mahdi et al., 2006; Sanchez-Iglesias et al., 2009; Sumathi et al., 2013). Although OS and neuron alterations in brain tissue of neonatal rats have been reported by direct exposure of the pups to Al (Yuan et al., 2011, 2012), information on such OS in developing brain of the pups whose mothers have been exposed to Al during gestation and lactation periods is lacking.

Thus, the effect of perinatal oral Al-administration to pregnant mice needs to be assessed in the offspring at neurobehavioral, cognitive and biochemical levels (brain neurotransmitters and OS indices) to investigate the possible long term physiological mechanism associated with Al toxicity related with cognitive dysfunction in the offspring. All data on maternal effects itself, have been excluded from this chapter, however; no effects were noticed on the dams.

MATERIALS AND METHODS

Animals

Male and female Swiss–Webster strain mice (10–12 weeks old) were housed in opaque plastic cages (three females to one male in each cage) measuring 30×12×11 cm, in the animal facility of the Zoology Department, King Saud University, Riyadh, Saudi Arabia.

Animals were kept under reversed lighting conditions with white lights on from 22.30 to 10.30 h local time. The ambient temperature was regulated between 18 and 22°C. After pregnancy (appearance of vaginal plug was considered as day one of pregnancy), the males were removed from the cages and the females were subjected to experimental treatments. Food (Pilsbury's Diet) and water were available ad libitum, unless otherwise indicated.

Experimental Design for Al Treatments

All pregnant mice were divided into 3 groups. The first group served as the control group and received plain tap water only, whereas the second and third groups were exposed to 300 and 600 mg/kg aluminum chloride (Al) per day respectively, dissolved in plain tap water, through oral administration. Our pilot studies have shown that the normal and/or pregnant mice on an average consume 30 ml water per day and thus the Al doses were prepared accordingly. These Al doses formed the sole drinking fluid source for the experimental group of pregnant mice during the required period of the experiment. It is likely that the present high concentrations of Al used in the drinking water might have affected total fluid intake due to its astringent property, but no efforts were made to assess any such differences in total fluid intake specifically. Fresh Al doses were replaced in the drinking bottle every day. The factor for the possibility of presence of Al traces in food and tap water was not taken into account for calculating the daily Al intake. However, this factor was minimized by giving the same source of food and tap water to all experimental groups including the controls. All pregnant mice were housed individually. Treatment started from day 1 of pregnancy, and was continued until postnatal day 15 (PD15) and thereafter the mothers were switched to plain tap water. The pups of each experimental group were culled to only eight per dam on the day of birth (PD0) and were left with their mothers until PD22. During this weaning period, three male pups of each litter were color marked from the others, and were subjected to various behavioral tests (described below) under dim lighting (ca 8 lux). In all, 21 pups belonging to seven litters from each treatment category were considered. All observations were recorded on PD1 and repeated every other day until PD21 in the same cohort of three color marked male pups of each litter. These observations were used to measure the early development of sensory motor coordination reflexes together with morphological development in the pups. For statistical analysis, the mean of all three cohorts (color marked pups) per litter was considered as a single score. Thus, seven replicates from each treatment category were considered for the following observations.

Physical Assessment during Weaning Period

Physical developmental landmarks like body weight, opening of the eyes and appearance of body hair fuzz, were evaluated in the developing offspring starting from day 1 after birth (PD1) through the entire weaning period until PD21.

Body Weight

Weight is a useful indicator of development. Thus, the pups were weighed every alternate day from PD1 until PD21.

Eye Opening and Hair Appearance

The day at which the body hair fuzz appeared, and the eyes opened was also recorded. These two parameters are also useful morphological indicators of development.

Neuromotor Maturation Assessment during Weaning Period

Righting Reflex

The time taken by a pup placed on its back to turn over and place all four paws on the substrate was recorded. An upper limit of 2 min was set for this test.

Cliff Avoidance

Pups were placed on the edge of a table top with the forepaws and face over the edge. The time taken by the pup to back away and turn from the "cliff" was recorded. Again an upper limit of 2 min was chosen. A latency of 2 min was attributed when the animal fell from the "cliff".

Rotating Reflex

The surface used to measure the rotating reflex was the same as that used for righting reflex, except that it was inclined at an angle of 30°. The pups were placed on this surface with their heads pointed downward. The time elapsed until the pup rotated its body through 180° geonegatively and faced its head upward, was recorded as the rotating time. The upper limit of this test was also set at 2 min.

Cognitive Behavioral Assessment

The following tests were evaluated in the same cohort of male offspring (bearing in mind to include representatives of each litter) in all the behavioral tests.

Active Avoidance Responses

The active avoidance responses were measured in the animals at PD 25, using an automatic reflex conditioner "shuttle box" (Ugo Basile, Comerio, Varese, Italy). The rectangular shaped shuttle-box was divided into two chambers of equal size by a stainless steel partition with a gate providing access to the adjacent compartment. Before starting the trial sessions, each animal was allowed to adapt and acquaint itself with the shuttle box for 2 min without any stimulus. A 6 s duration light (21W) and buzzer (670 Hz and 70 dB) were switched on consecutively and used as a conditioned stimulus (CS). The CS preceded the onset of the unconditioned stimulus (US) by 5 s. The US was an electric scrambler shock (1 mA for 4 s) applied to the grid floor. If the animal avoided the US by running into the other compartment within 5 s after the onset of the CS, the microprocessor recorder unit of the shuttle box recorded an avoidance response and this was considered as conditioned avoidance response to avoid the electric shock. Each animal was given 50 trials with a fixed intertrial interval of 15 s. During the 50 trial session of the individual animal, the total number of avoidance was measured. The total time taken until the animal entered the other compartment to avoid the shock treatment (latency of avoidance response or escape latency in seconds) was also measured for each animal and the results were expressed for each group of animals. The spontaneous migration of the mouse to the other compartment between trials was also assessed by measuring the number of crossings between the chambers when no shock was present during UCS and CS (inter-trial crossing). The recorder unit of the automated shuttle box continuously recorded these parameters during the whole experimental period (50 trials) of each animal.

Morris Water-Maze Test

The test has been extensively used to assess cognitive functions in rat (Rutten et al., 2002; Tariq et al., 2008) and mice (Lamberty and Gower, 1991a, 1991b) models. Starting at the age of PD30, the mice offspring were tested for visual–spatial memory using a water-maze (Morris, 1984). The water-maze consisted of a galvanized white circular water tank (90 cm diameter, 50 cm height) filled with clear tap water (22 ± 1°C) to a depth of 15 cm. A 6×6 cm size, stainless steel, adjustable, white, escape platform was placed 1 cm below the water level and 13 cm from the rim. The water was made opaque by addition of 1 l of milk, which prevented visualization of the platform. Four points on the rim of the tank were designated north (N), south (S), east (E) and west (W), thus dividing the pool into four quadrants (NW, NE, SE and SW). On the first day, each offspring (P30) was allowed to swim freely in the pool for 60 s without the platform present in the pool. This free swim enabled the mice to become habituated to the training environment. On days 2–5, offspring (P31 to P35, respectively) were trained for 24 trials (six trials a day, with an inter-trial interval of 30 s) to locate and escape onto the submerged platform. At the start of each trial, the mouse was held facing the perimeter of the water tank and dropped into the pool to ensure immersion. The latency from immersion into the pool to escape onto the hidden platform (maximum duration of trial 120 s) was recorded. If the mouse did not find the platform in 120 s, it was manually guided with the help of a glass rod to mount on the platform and a score of 120 s was recorded for each of such experimenter-terminated trials. The number of such unsuccessful trials was counted and expressed as a percentage of failures on each testing day. On mounting the platform, each mouse was given a 30 s inter-trial interval for rest and for learning and memorizing the spatial cues to reach the platform for escape. To minimize handling, at the end of the trials, the animals were allowed to climb onto a wire mesh grid and transferred to their cage without further handling.

On day 6, P 36 mice were subjected to a 120 s probe trial in which the platform was removed from the pool. The time spent in each quadrant (within 120 s probe test time) was recorded on an electronic time recorder. In this part of probe trials in water-maze test, normal animals typically spend more time in the quadrant where the platform had been previously located than in the other quadrants. The testing procedures used during the four days of locating the hidden platform provide a measure of hippocampal-dependent spatial reference memory, while the probe trial is a measure of the strength of spatial learning, the closest parallel to episodic memory in humans (Jeltsch et al., 2001; Spiers et al., 2001).

Biochemical Studies

For biochemical studies, a sub-set of developing offspring (n = 8/group) were sacrificed at different ages (PD7, 14, 21, 30 and 36) by decapitation and their brains were dissected on ice. The fore-brain was isolated (including the cerebral areas with hippocampus and striatum) and frozen in liquid nitrogen and stored at −70°C for neurotransmitters and OS indices estimations.

Determination of Monoamines

The monoamines, dopamine (DA) and serotonin or 5-hydroxytryptamine (5-HT) were estimated using the modified method of Patrick et al. (1991). A 10% homogenate of the fore-brain was prepared by homogenizing the tissues for 10 s in 0.1 M HClO4 containing 0.05% EDTA, centrifuged at 17,000 rpm at 4°C for 5 min. The supernatants were filtered using 0.45 µm pore filters and analyzed by high performance liquid chromatography (HPLC). The mobile phase consisted of 32 mM citric acid monohydrate, 12.5 mM disodium hydrogen orthophosphate, 7% methanol, 1 mM octane sulfonic acid and 0.05 mM EDTA. The mobile phase was filtered through 0.22 µm filter and degassed under vacuum before use. µBondpak C18 column was used at a flow rate of 1.2 ml/min and the injection volume of the sample was 20 µl. The levels of DA and 5-HT were calculated using a calibration curve and results were expressed as ng/mg tissue weight.

Determination of Nonenzymatic OS Indices

Lipid Peroxides

Lipid peroxides (LP) were determined spectrophotometrically as thiobarbituric acid-reactive substances (TBARS) according to the method of Ohkawa et al. (1979). Tissue lipid peroxide levels were quantified using extinction coefficient of 1.56×10^5 m−1 cm−1 and expressed as nanomoles of TBARS formed per g tissue weight. The results are expressed as nmol/g wet weight.

Glutathione

Reduced glutathione (GSH) level was measured enzymatically in the brain tissue by a slightly modified method (Mangino et al., 1991). The slope of the change in absorbance was used to quantitate total GSH by comparing the slope of the samples with a standard curve prepared with pure glutathione (Sigma). The specific activity is expressed into umol/g tissue weight.

Determination of Enzymatic OS Indices

Glutathione-S-Transferase

Glutathione S-transferase (GST) was estimated by the method of Habig et al. (1974) by using 1-chloro-2,4-dinitrochlorobenzene (CDNB) as substrate at 340 nm. The GST activity is expressed as U/g tissue weight.

Catalase

Catalase (CAT) activity was measured by the method of Aebi (1972), by tracking the decomposition of hydrogen peroxide by measuring decrease in extinction of H_2O_2 at 240 nm. The activity of CAT is expressed as rate constant of first order reaction K per gram tissue weight.

Superoxide Dismutase

Superoxide dismutase (SOD) activity was estimated by the method of Misra and Fridovich (1972). Activity is expressed as the amount of enzyme that inhibits the oxidation of epinephrine by 50% which is equal to U per gram tissue weight.

Statistical Analysis

All data were analyzed by using the Student–Newman–Keuls multiple comparison test of ANOVA by the INSTAT computer program. Significant effects in locomotory test were further evaluated using Dunn's multiple comparison tests.

RESULTS

Physical Assessment during Weaning Period

Physical developmental landmarks were evaluated in the developing offspring starting from the day 1 after birth (PD1) through entire weaning period unto PD21. Although the body weights of pups born to Al-treated mice lagged behind the control group on PD1, the difference was statistically insignificant. However, the Al exposed offspring at both doses, showed a significant decline ($p < 0.05$) in their body weight gain as compared to the controls, from PD3 onwards. From PD5 upto the weaning period (PD21), the Al exposed offspring remained lagging behind the controls in body weight very significantly ($p < 0.001$) in a dose-dependent manner (Figure 1).

Other morphological developments such as the opening of the eyes and appearance of body hair fuzz were also significantly ($p < 0.001$) delayed in the Al exposed offspring in a dose-dependent manner (Figure 2).

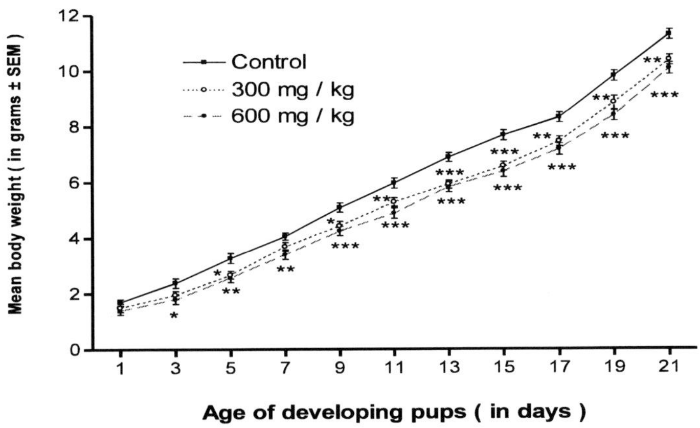

Figure 1. Effect of perinatal aluminium doses (300 and 600 mg/kg body weight) exposure on the body weight gain of mouse pups during the weaning (lactation) period. (*, **and ***) represent statistically significant (*P < 0.05, P < 0.01 and P < 0.001 respectively)* from the control group; see text.
(Taken from: Abu-Taweel et al., 2012 *Pharmacology, Biochemistry and Behavior*, 101: 49 – 56.)

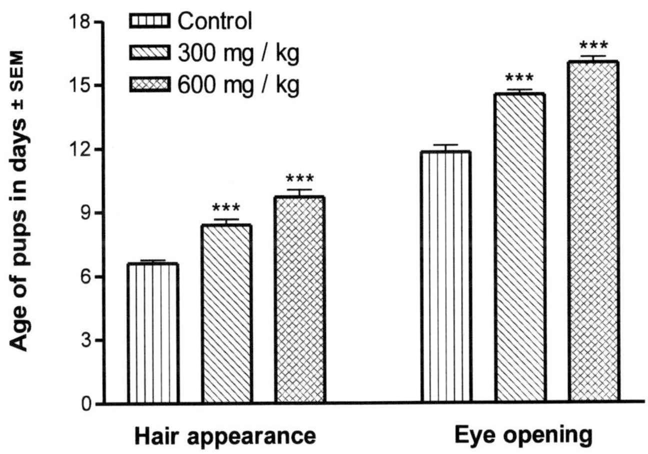

Figure 2. Effect of perinatal aluminium doses (300 and 600 mg/kg body weight) exposure on the body hair appearance and eye opening in the mouse pups. (***) represents statistically significant (*P < 0.001)* from the control group; see text.
(Taken from: Abu-Taweel et al., 2012 *Pharmacology, Biochemistry and Behavior*, 101: 49 – 56.)

Behavioral Studies

Neuromotor Maturation

The neuromaturation of reflexes during the weaning period of the developing pups was assessed from the day of birth PD1 until PD21. The righting, rotating, and cliff avoidance reflexes in the Al-exposed offspring were found to be significantly and dose-dependently suppressed throughout the weaning period (Figure 3).

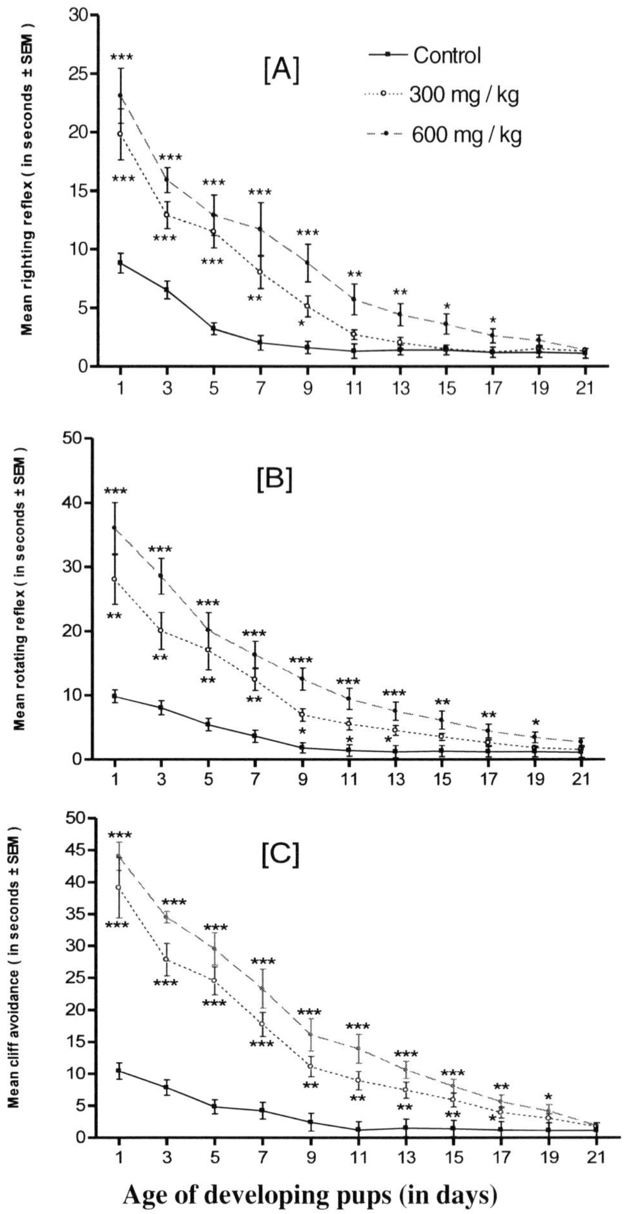

Figure 3 A – C. Effect of perinatal aluminium doses (300 and 600 mg/kg body weight) exposure on the righting reflex (A), rotating reflex (B) and cliff avoidance (C) of mouse pups during the weaning (lactation) period. (*, **and ***) represent statistically significant (*P < 0.05, P < 0.01 and P < 0.001 respectively)* from the control group; see text.
(Taken from: Abu-Taweel et al., 2012 *Pharmacology, Biochemistry and Behavior*, 101: 49 – 56.)

Active Avoidance Test

In the shuttle-box active avoidance test, the Al-exposed offspring (PD25), showed a statistically significant and dose-dependent decrease in the number of avoidances during the trial period as compared to the control group (Figure 4 A). The spontaneous migration of the mouse to the other compartment between trials measuring the number of crossings between the chambers when no shock was present (inter-trial crossing) showed a significant and dose-dependent decrease in the number of inter-trial crossings as compared to the controls (Figure 4B). The total time taken during the entire trials by the animals to enter the other compartment to avoid the shock treatment (latency of avoidance response in seconds) was also measured for each animal. The results showed that the animals exposed to Al were poor learners in a dose-dependent manner and took significant time in avoiding the shock treatment as compared to the controls (Figure 4 C).

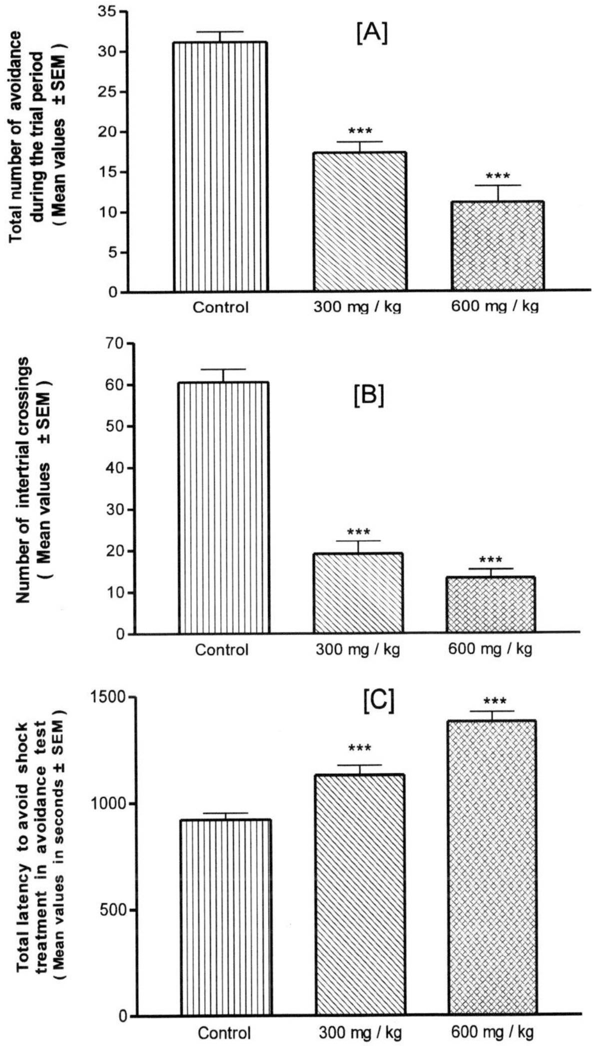

Figure 4 A – C. Effect of perinatal aluminium doses (300 and 600 mg/kg body weight) exposure on the mean performance value of the mice offspring at the age of 25 days postnatal in the active avoidance task. Mice were given a 50 trial session and the total number of times they avoided the shock by moving to the other compartment of the shuttle box (A), the number of inter-trial crossings between the chambers in the absence of current shock (B) and the total time taken by the animals (latency) to avoid the shock (C) were measured. (***) represents statistically significant ($P < 0.001$) from the control group; see text.
(Taken from: Abu-Taweel et al., 2012 *Pharmacology, Biochemistry and Behavior*, 101: 49 – 56.)

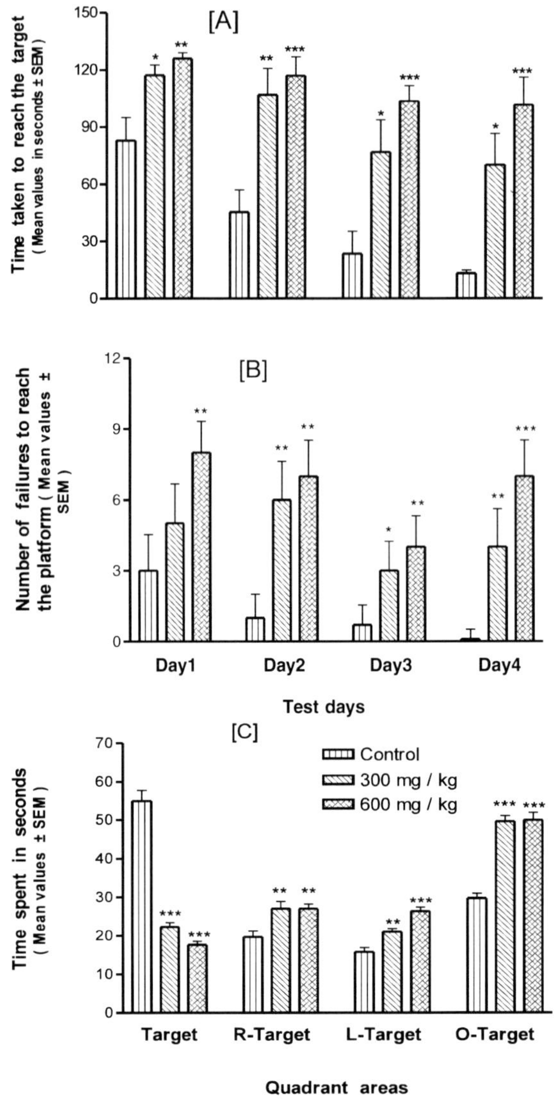

Figure 5 A – C. Performance in Morris water-maze of mice offspring at the age of 30 days postnatal whose mothers were exposed perinatally with aluminium doses of 300 and 600 mg/kg body weight.
A – shows the mean latency to reach the hidden platform (y-axis) on each testing day (x-axis). Animals subjected to aluminium exposure were slower in finding the platform than the controls on all testing days.
B – shows the number of failures or unsuccessful trials (y-axis) on each testing day (x-axis). Aluminium exposed offspring showed maximum number of failures in finding the platform as compared to controls.
C – shows probe test performance in Morris water-maze of mice offspring at the age of 36 days. The aluminium exposed offspring spent less time in the target quadrant than the control group. (*, **and ***) represent statistically significant (*$P < 0.05$, $P < 0.01$ and $P < 0.001$ respectively*) from the control group; see text. R-Target denotes quadrant on the right side of the target quadrant, L-Target denotes the quadrant on the left side of the target quadrant and O-Target represents quadrant opposite to the target quadrant.
(Taken from: Abu-Taweel et al., 2012 *Pharmacology, Biochemistry and Behavior*, 101: 49 – 56.)

Morris Water-Maze Task

Mice offspring with Al treatment, exhibited longer escape latencies to reach the platform as compared with the control group ($p < 0.01$; Figure 5 A), however, all groups displayed a gradual improvement in performance over the 4 days of testing (training) period. The number of unsuccessful trials (failures) to reach the platform was also significantly higher in Al treated offspring as compared to the control group on all testing days ($p < 0.001$; Figure 5 B).

The probe trial studies showed that Al exposed offspring spent more time in other three quadrants than the target (platform) quadrant as compared to the control group ($p < 0.001$; Figure 5 C), in search of the platform.

Biochemical Studies

Levels of Monoamines in Forebrain

There was a significant ($p < 0.001$) and dose-dependent inhibition of dopamine level in the forebrain of mice offspring treated with Al as compared to the control group at the ages PD7, PD14, PD21, PD30 and PD36 (Figure 6 A). On the contrary, the level of 5-HT was significantly ($p < .001$) depleted in the offspring at all developing ages, exposed only to higher dose of Al, whereas, the lower dose had no effect at any developing ages (Figure 6 B).

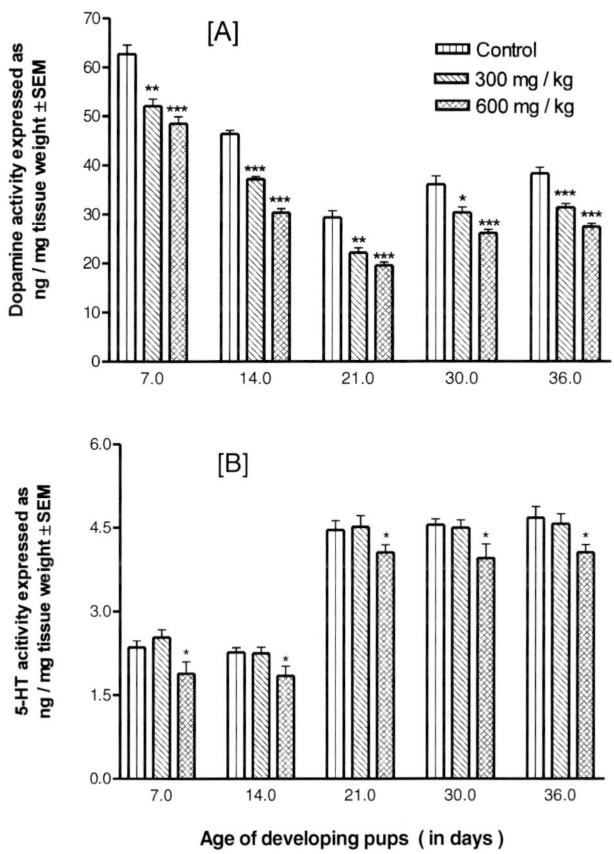

Figure 6 A and B. Effect of perinatal aluminium doses (300 and 600 mg/kg body weight) exposure on the mean levels of dopamine (A) and 5-HT (5-hydroxy-tryptamine or serotonin) (B), in the forebrain of the offspring at various postnatal developing ages (x-axis). (*, **and ***) represent statistically significant (*$P < 0.05$, $P < 0.01$ and $P < 0.001$ respectively*) from the control group; see text. (Taken from: Abu-Taweel et al., 2012 *Pharmacology, Biochemistry and Behavior*, 101: 49 – 56.)

Levels of Nonenzymatic OS Indices

LP determined as TBARS were found to be elevated significantly (p < 0.001) due to perinatal Al exposure in the developing forebrain of the offspring throughout the postnatal development period (PD7, PD14 and PD21) and even at adolescent ages PD30 and PD36 in a dose-dependent manner (Figure 7 A). On the contrary, reduced glutathione (GSH) level remained depleted significantly (p < 0.001) at all developing ages in a dose-dependent manner (Figure 7 B).

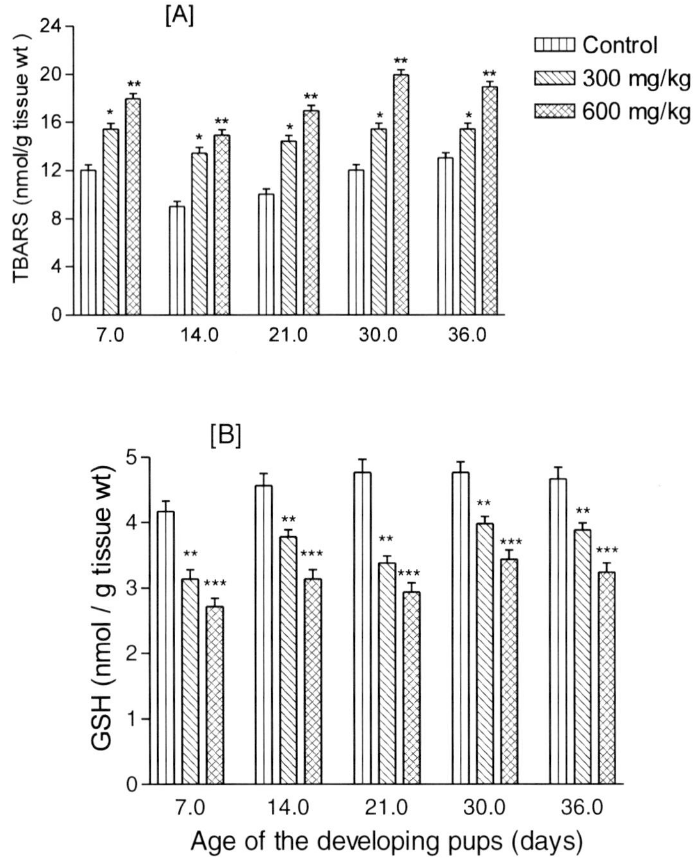

Figure 7 A and B. Effect of perinatal aluminium doses (300 and 600 mg/kg body weight) exposure on the mean levels of nonenzymatic oxidative stress indices like (A) lipid peroxidation content (TBARS), and (B) total glutathione (GSH) level, in the forebrain of the offspring at various postnatal developing ages (x-axis). (*, **and ***) represent statistically significant (*P < 0.05, P < 0.01 and P < 0.001 respectively)* from the control group; see text.

Levels of Enzymatic OS Indices

The levels of enzymatic OS indices GST, CAT, and SOD remained significantly (p < 0.001) depleted in a dose-dependent manner in the fore-brain of the developing offspring at PD7, PD14, PD21, PD30 and PD36 ages (Figures 8 A, B, and C respectively).

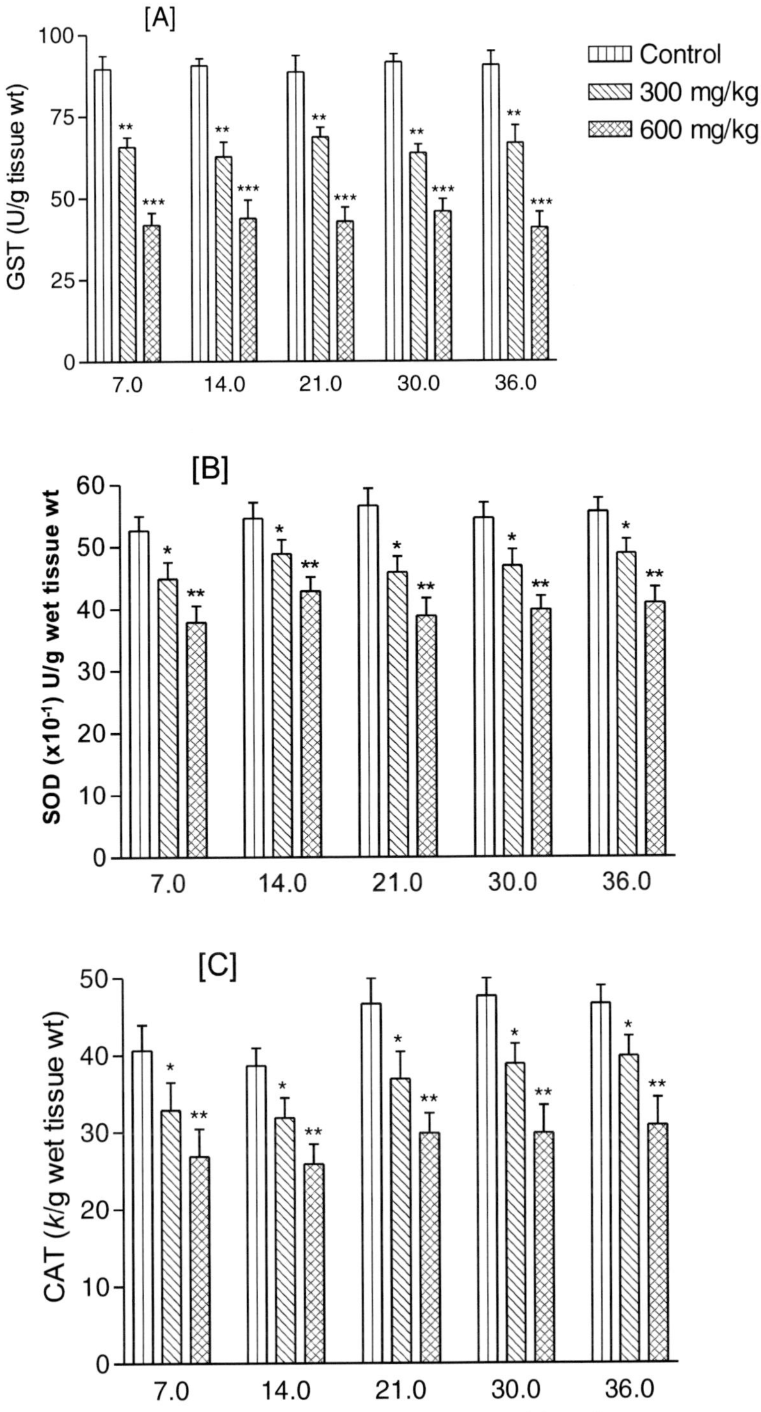

Figure 8 A – C. Effect of perinatal aluminium doses (300 and 600 mg/kg body weight) exposure on the mean levels of enzymatic oxidative stress indices like (A) glutathione S-transferase (GST) level, (B) catalase (CAT) activity, and (C) superoxide dismutase (SOD) activity in the forebrain of the offspring at various postnatal developing ages (x-axis). (*, **and ***) represent statistically significant (*$P < 0.05$, $P < 0.01$ and $P < 0.001$ respectively)* from the control group; see text.

DISCUSSION

The present results demonstrate that female mice exposed to Al during pregnancy produce pups that markedly differ from their controls in the rate of physical maturation, neuromotor development, cognitive responses in the active avoidance test and Morris water-maze test, in the levels of neurotransmitters and oxidative stress indices in their forebrain region. The postnatal suppression of body weight gain and delay in opening of eyes and the appearance of body hair fuzz in the Al exposed pups, might indicate a lasting effect of the perinatal Al exposure on general growth retardation in mice offspring. Perinatal Al administration has also affected the preweaning reflexes in the mice pups. All reflexes including righting, rotating and avoiding of the cliff by the developing pups were significantly suppressed as compared to the controls. This clearly suggests for a direct Al intervention with the developing pups in utero because Al has been shown to be transferred to offspring through placenta or milk (Abu-Taweel et al., 2012). A recent study using radio isotopic Al showed that considerable amounts of Al administered to pregnant and/or lactating rats, crossed the blood brain barrier and was deposited into the brain of fetuses and sucklings through the transplacental passage and/or maternal milk which remained persistent throughout their lifetime (Yumoto et al., 2001). Thus, Al exposure during fetal life does retard motor development and physical maturation, as have been suggested for other drugs (Brain et al., 1994) and compounds (Ajarem and Ahmad, 1991; 1998).

It is well established that Al is toxic to the growth and development of fetuses and sucklings in humans and experimental animals (Domingo, 1995; Yumoto et al., 2001; Sharma and Mishra, 2006). Gestational and lactational exposure to Al doses can result in persistent neurobehavioural deficits in the offspring of some mammals (Roig et al., 2006). Studies in rodents have also demostrated that during pregnancy, exposure to Al causes long-lasting effects on emotionality and learning capabilities (Nishio et al., 2001; Sternberg and Ridgway, 2003; Shaw and Petrik, 2009; Kumar et al., 2009). Al exposure in animals and humans results in behavioral changes and intellectual impairment and it is possible that the behavioral changes could be the result of subtle changes in the serotonin level of the brain regions following Al exposure (Kumar, 2002).

The results of this study showed that the levels of dopamine was reduced significantly and dose-dependently whereas 5-HT was inhibited only at the higher dose of Al in the forebrain tissue of the Al-exposed mice offspring at all ages. There is evidence of an inhibitory role of DA mediated receptor (D_2 type) in depressing the hyperexcitability of hippocampal and striatal neurons (Freitas et al., 2004; Nascimento et al., 2005). A number of 5-HT receptor subtypes have been reported for having different roles in the functions of serotonergic neurotransmission, including the functions connected with learning and memory processes (Petkov et al., 1995). The mice offspring exposed to Al tended to perform badly in water maze parameters and also resulted in decreased number of avoidances in the automatic reflex conditioner as compared to the control offspring. This suggests for a tendency towards decreasing of the memory effect of Al under conditions of reduced functional capacity of serotonergic neurotransmission. Recently, a growing body of research has focused on the participation of serotonin (5-HT) in the neurochemical mechanisms of cognition and especially of learning and memory. Potential toxic mechanisms of action for Al may include disruption in serotonergic neurotransmission through disturbed levels of neurotransmitters in the brain hippocampus (Richter-Levin and Segal, 1991). Other reported mechanisms of action of Al toxicity may include enhancement of inflammation (i.e., microgliosis) and the interference with cholinergic projections (Platt et al., 2001), reduced glucose utilization (Joshi. 1990), defective phosphorylation-dephosphorylation reactions (Cordeiro et al., 2003), altered rate of transmembrane diffusion and selective changes in saturable transport systems in the blood brain barrier (Kaya et al., 2003), oxidative damage on cellular processes by the inhibition of the glutathione redox cycle (Murakami, 1999) and inflicting damage to membrane lipid, proteins and antioxidative enzyme defense system (Jyoti et al., 2007). Al has been reported to alter blood-brain barrier (Banks and Kastin, 1985; Zatta et al., 2002) and gets deposited in the cortex, cingulate bundles, corpus callosum (Plat et al., 2001) and hippocampus (Struys-Ponsar et al., 1997; Fattoretti et al., 2004). As an important target organ of neurotoxicity, the hippocampus (located in the forebrain region of rodents) is a crucial element of the neurotoxicity basis of higher cognitive function (Savage et al., 2004; Tariq et al., 2008). It is evidently suggested in the present chapter that the brain may be the most susceptible target organ for Al toxicity.

The present chapter indicated a significant and dose-dependent increase in TBARS and decrease in GSH and CAT, GST, and SOD levels in the forebrain region of the mice treated with Al clearly suggesting for a high level of enzymatic (GST, SOD, and CAT) and nonenzymatic (TBARS and GSH) OS in a long lasting manner. It is documented that in infants although their growth rate is rapid, their blood brain barrier, detoxification system (liver), and excretory system (kidney) are not well developed (Snell, 2001; Blackburn and Loper, 1992). Crossing of Al through the blood brain barrier and accumulation in the brain glial and neural cellsg and dose dependent OS in the forebrain of the mice offspring inflicting cognitive dysfunctions.

CONCLUSION

From the present results it can be concluded that Al exposure during pregnancy can affect the fetus, raising the concerns that during a critical perinatal period of brain development, Al exposure might modify the properties of the dopaminergic system and thus change the threshold of that system or other related systems. Furthermore, Al exposure during perinatal period may disrupt the steady state balance between ROS production (TBARS) and their scavenging by the cellular antioxidant system (GSH). Perinatal Al exposure alters the brain enzymatic (GST, CAT, SOD) and nonenzymatic (TBARS and GSH) antioxidant defense system (AOS) and causes a long lasting and dose dependent effects. Disruption of neuromotor reflexes in the neonatal stage and cognitive dysfunction evidenced through shuttle box and water maze tests in the offspring at adolescent ages further support the possibility that Al exposure during pregnancy (perinatal period) has potential neurotoxic hazards to the in utero developing fetus brain. Reduced use of Al during pregnancy is of crucial importance in preventing Al-induced neurotoxicity in the offspring. In addition, a biogenetical level of investigations in context to perinatal Al exposure in offspring for a longer lasting effects may play a crucial role and may be a potential area for further studies.

ACKNOWLEDGMENT

The authors are thankful to the Deanship of Scientific Research, College of Nursing Research Center at King Saud University for funding this research.

REFERENCES

Abu-Taweel, G.M., Ajarem, J.S. and Ahmad, M. 2012. Neurobehavioral toxic effects of perinatal oral exposure to aluminum on the developmental motor reflexes, learning, memory and brain neurotransmitters of mice offspring. *Pharmacology, Biochemistry and Behavior* 101: 49 – 56.

Aebi, H. 1972. "Catalase." In *Methods of Enzymatic Analysis*, edited by H. U. Bergmeyer, New York: Academic Press.

Ajarem, J.S. and Ahmad, M. 1991. Behavioral and biochemical consequences of perinatal exposure of mice to instant coffee: a correlative evaluation. *Pharmacology Biochemistry and Behavior* 40: 847 – 852.

Ajarem, J.S. and Ahmad, M. 1998. Prenatal nicotine exposure modifies behavior of mice through early development. *Pharmacology Biochemistry and Behavior* 59: 313 – 318.

Banks, W.A. and Kastin, A.J. 1985. The aluminum-induced increase in the blood-brain barier permeability to delta-sleep-inducing peptide occurring through the brain is independent of phosphorus and acetylcholinesterase levels. *Psychopharmacology* 86: 84 – 89.

Bishop, N.J., Morley, R., Chir, B., Day, J.P. and Lukas, A. 1997. Aluminum neurotoxicity in preterm infants receiving intravenous feeding solutions. New England Journal of Medicine 336: 1557 – 1561.

Blackburn, S.T. and Loper, D.L. 1992. *Maternal, fetal, and neonatal physiology, a clinical perspective.* Philadelphia: W. B. Saunders Company.

Brain, P.F., Kurishingal, H., Whiting, C.J. and Restall, C.J. 1994. An Ethopharmacological Approach to Behavioral Teratology. In *Ethology and Psychopharmacology*, edited by S.J. Cooper, and C.A. Hendrie, 224 – 239, John Wiley and Sons Ltd., New York.

Colomina, M.T., Sanchez, D.J., Domino, J.L. and Sanchez-Turet, M. 1999. Exposure of pregnant mice to aluminum and restraint stress: effects on postnatal development and behavior of the offspring. *Psychobiology* 27: 521 – 529.

Colomina, M.T., Roig, J.L., Torrente, M., Vicens, P. and Domingo, J.L. 2005. Concurrent exposure to aluminum and stress during pregnancy in rats: effects on postnatal development and behavior of the offspring. *Neurotoxicology and Teratology* 27:565 – 574.

Cordeiro, J.M., Silva, V.S., Oliveira, C.R. and Goncalves, P.P. 2003. Aluminium-induced impairment of Ca^{2+} modulatory action on GABA transport in brain cortex nerve terminals. *Journal of Inorganic Biochemistry* 97: 132 – 142.

Domingo, J.L. 1995. Reproductive and developmental toxicity of aluminum: a review. *Neurotoxicology and Teratology* 17: 515 – 521.

Exley, C. 2001. Aluminium and Alzheimer's disease. *Journal of Alzheimers Disease* 3: 551 - 552.

Fattoretti, P., Bertoni-Freddari, C., Balietti, M., Georgetti, B., Solazzi, M. and Zatta, P. 2004. Chronic aluminum administration to old rats results in increased levels of brain metal ions and enlarged hippocampal mossy fibers. *Annales of New York Academy of Sciences* 1019: 44 – 47.

Forbes, W.F., Gentleman, J.F. and Maxwell, C.J. 1995. Concerning the role of aluminum in causing dementia. *Experimental Gerentology* 30: 23 – 32.

Freitas, R.M., Vasconcelos, S.M.M., Souza, F.C.F., Viana, G.S.B. and Fonteles, M.M.F. 2004. Monoamine levels after pilocarpine-induced status epilepticus in hippocampus and frontal cortex of Wistar rats. *Neuroscience Letters* 370: 196 – 200.

Golub, M.S. and Domingo, J.L. 1996. What we know what we need to know about developmental aluminum toxicity. *Journal of Toxicology and Environmental Health* 48: 585 – 597.

Golub, M.S. and Germann, S.L. 2001. Long-term consequences of developmental exposure to aluminum in a suboptimal diet for growth and behavior of Swiss Webster mice. *Neurological Toxicologo* 23: 365 – 372.

Golub, M.S., Han, B. and Keen, C.L. 1999. "Aluminium uptake and effects on transferrin mediated iron uptake in primary cultures of rat neurons, astrocytes and oligodendrocytes." *Neurotoxicology* 20: 961 – 970.

Gonda, Z. and Lehotzky, K. 1996. Effect of prenatal aluminium lactate exposure on conditioned taste aversion and passive avoidance task in the rat. *Journal of Applied Toxicology* 16: 529 – 532.

Gonda, Z., Miklosi, A. and Lehotzky, K. 1997. The effect of social learning on a conditioned avoidance response of rats treated prenatally with aluminum lactate. *Neurotoxicology and Teratology* 19: 59 – 63.

Goodlet, C.R. and Horn, K.A. 2001. Mechanisms of alcohol-induced damage to the developing nervous system. *Alcohol Research and Health* 25: 175 – 184.

Greger, J.L. 1993. Aluminum metabolism. *Annual Review in Nutrition* 13: 43 – 63.

Habig, W.H., Pabst, M.J. and Jakoby, W.B. 1974. "Glutathione S-transferases. The first enzymatic step in mercapturic acid formation." *The Journal of Biological Chemistry* 249: 7130–7139.

Jeltsch, H., Bertrand, F., Lazarus, C. and Cassel, J.C. 2001. Cognitive performances and locomotor activity following dentate granule cell damage in rats: role of lesion extent and type of memory tested. *Neurobiology of Learning and Memory* 76: 81 – 105.

Joshi, J.G. 1990. Aluminum, a neurotoxin which affects diverse metabolic reactions. *Biofactors* 2: 163 – 169.

Jyoti, A., Sethi, P. and Sharma, D. 2007. Bacopa moniera prevents from aluminium neurotoxicity in the cerebral cortex of rat brain. *Journal of Ethnopharmacology* 111: 56 – 62.

Kawahara, M. 2005. Effects of aluminum on the nervous system and its possible link with neurodegenerative diseases. *Journal of Alzheimers Disease* 8: 171 – 182.

Kawahara, M. and Kato-Negishi, M. 2011. Link between aluminium and the pathogenesis of Alzheimer's disease: the integration of the aluminium and amyloid cascade hypotheses. *International Journal of Alzheimers Disease* 276393: 1 – 17.

Kaya, M., Kalayci, R., Arican, N., Kucuk, M. and Elmas, I. 2003. Effect of aluminum on the blood-brain barrier permeability during nitric oxide-blockade-induced chronic hypertension in rats. *Biological Trace Elements Research* 92: 221 – 230.

Kumar, S. 2002. Aluminium-induced changes in the rat brain serotonin system. *Food and Chemical Toxicology* 40: 1875 – 1880.

Kumar, A., Dogra, S. and Prakash, A. 2009. Protective effect of curcumin (Curcuma longa), against aluminium toxicity: Possible behavioral and biochemical alterations in rats. *Behvioral Brain Research* 205: 384 – 390.

Lamberty, Y. and Gower, A.J. 1991a. Simplifying environmental cues in a Morris-type water maze improves place learning in old NMRI mice. *Behavioral Neural Biology* 56: 89 – 100.

Lamberty, Y. and Gower, A.J. 1991b. Cholinergic modulation of spatial learning in mice in a Morris-type water maze. *Archives of Pharmacodynamics and Therapeutics* 309: 5 – 19.

Mahdi, A.A., Tripathi, S., Neerja, J. and Hasan, M. 2006. "Aluminium mediated oxidative stress: possible relationship to cognitive impairment of Alzheimers type." *Annals of Neurosciences* 13: 1 – 7.

Mangino, M.J., Murphy, M.K. and Glabar, G.G. 1991. "Protective effects of glycine during hypothermic renal ischemic reperfusion injury." *American Journal of Physiology* 261: F841 – F848.

Misra, H.P. and Fridovich, I. 1972. "The role of superoxide anion in the autoxidation of epinephrine and a simple assay for superoxide dismutase." *The Journal of Biological Chemistry* 247: 3170–3175.

Morris, R.G.M. 1984. Developments of a water-maze procedure for studying spatial learning in the rats. *Journal of Neuroscience Methods* 11: 47 – 60.

Murakami, N. 1999. Parkinsonism-dementia complex on Guam – overview of clinical aspects. *Journal of Neurology* 246: 16 – 18.

Nascimento, V.S., Oliveira, A.A., Freitas, R.M., Sousa, F.C., Vasconcelos, S.M.M., Viana, G.S.B. and Fonteles, M.M.F. 2005. Pilocarpine-induced status epilepticus: Monoamine level, muscuranic and dopaminergic receptors alterations in striatum of young rats. *Neuroscience Letters* 383: 165 – 170.

Nishio, H., Kasunga, S., Ushijima, M. and Harada, Y. 2001. Prenatal stress and postnatal development of neonatal rats – sex-dependent effects on emotional behavior and learning ability of neonatal rats. *International Journal of Developmental Neuroscience* 19: 37 – 45.

Ohkawa, H., Ohishi, N. and Tgi, K. 1979. "Assay for lipid peroxides in animal tissues by thiobarbituric acid reaction." *Annals of Chemistry* 95: 351 – 358.

Patel, A.J. 1983. "Undernutrition and brain development." *Trends in Natural Sciences* 6: 151 – 154.

Patrick, O.E., Hirohisa, M., Masahira, K. and Koreaki, M. 1991. Central nervous system bioaminergic responses to mechanic trauma. *Surgical Neurology* 35: 273 – 279.

Petkov, V.D., Belcheva, S., Konstantinova, E. and Kehayov, R. 1995. Participation of different 5-HT receptors in the memory process in rats and its modulation by the serotonin depleter p-chlorophenylalanine. *Acta Neurobiologiae Experimentalis* 55: 243 – 252.

Platt, B., Fiddler, G., Riedel, G. and Henderson, Z. 2001. Aluminium toxicity in the rat brain: Histochemical and immunocytochemical evidence. *Brain Research Bulletin* 55: 257 – 267.

Reuhl, K.R. 1991. "Delayed expression of neurotoxicity: the problem of silent damage." *Neurotoxicology* 12: 341 – 346.

Richter-Levin, G. and Segal, M. 1991. The effects of serotonin depletion and raphe grafts on hippocampal electrophysiology and behavior. *Journal of Neurosciences* 11: 1585 – 1596.

Roig, J.L., Fuentes, S., Colomina, M.T., Vicens, P. and Domingo, J.L. 2006. Aluminum, restraint stress and aging: Behavioral effects in rats after 1 and 2 years of aluminum exposure. *Toxicology* 218: 112 – 124.

Rutten, A., van Albada, M., Silveira, D.C., Cha, B.H., Liu, X., Hu, Y.N., Cilio, M.R. and Holmes, G.L. 2002. Memory impairment following status epilepticus in immature rats: time-course and environmental effects. *European Journal of Neuroscience* 16: 501 – 513.

Sanchez-Iglesias, S., Mendez-Alvarez, E., Iglesis-Gonzalez, J., Munoz-Patino, A., Sanchez-Sellero, I., Labandeira-Garcia, J.L. and Soto-Otero, R. 2009. "Brain oxidative stress and selective behavior of aluminium in specific areas of rat brain: potential effects in a 6-OHDA-induced model of Parkinson's disease." *Journal of Neurochemitry* 109: 879 – 888.

Savage, L.M., Buzzetti, R.A. and Ramirez, D.R. 2004. The effects of hippocampal lesions on learning, memory, and reward expectancies. *Neurobiology of Learning and Memory* 82: 109 – 119.

Sethi, P., Jyoti, A., Singh, R., Hussain, E. and Sharma, D. 2008. Aluminium-induced electrophysiological, biochemical and cognitive modifications in the hippocampus of aging rats. *Neurotoxicology* 29: 1069 – 1079.

Sharma, P. and Mishra, K.P. 2006. Aluminum-induced maternal and developmental toxicity and oxidative stress in rat brain: Response to combined administration of Tiron and glutathione. *Reproductive Toxicology* 21: 313 – 321.

Shaw, C.A. and Petrik, M.S. 2009. Aluminum hydroxide injections lead to motor deficits and motor neuron degeneration. *Journal of Inorganic Biochemistry* 103: 1555 – 1562.

Snell, R.S. 2001. *Clinical neuroanatomy for medical students (Periodicals).* 5th Edition. Philadelphia: Lippincott Williams and Wilkins.

Spear, L.P. and File, S.E. 1996. "Methodological considerations in neurobehavioral teratology." *Pharmacology, Biochemistry and Behavior* 55: 455 – 457.

Spiers, H.J., Burgess, N., Hartley, T., Vargha-Khadem, F. and O'Keefe, J. 2001. Bilateral hippocampal pathology impairs topographical band episodic memory but not visual pattern matching. *Hippocampus* 11: 715 – 725.

Sternberg, W.F. and Ridgway, C.G. 2003. Effects of gestational stress and neonatal handling on pain, analgesia, and stress behavior of adult mice. *Physiolgy and Behavior* 78: 375 – 383.

Struys-Ponsar, C., Kerkhofs, A., Gauthier, A., Soffie, A.M. and van den Bosch. 1997. Effects of aluminum exposure on behavioral parameters in the rats. *Pharmacology Biochemistry and Behavior* 56: 643 – 648.

Sumathi, T., Shobana, C., Mahalakshmi, V., Surekha, R., Subathra, M., Vishali, A. and Rekha, K. 2013, "Oxidative stress in brains of male rats intoxicated with aluminium and neuromodulating effect of Celastrus paniculatus alcoholic seed extract." 6: 80 – 90.

Tariq, M., Ahmad, M., Moutaery, K.A. and Deeb, S.A. 2008. Pentoxifylline ameliorates lithium-pilocarpine induced status epilepticus in young rats. *Epilepsy and Behavior* 12: 354 – 365.

Walton, J.R. 2007. A longitudinal study of rats chronically exposed to aluminum at human dietary levels. *Neuroscience Letters* 412: 29 – 33.

World Health Organization. 2006. Aluminum. In: 657, Aluminum (WHO Food Additives Series 24) online, http://www.inchem.org/documents/jecfa/jeemono/v024je07.htm.

Yokel, R.A. 2000. The toxicology of aluminum in the brain: a review. *Neurotoxicology* 21: 813 – 828.

Yasui, M., Kihira, T. and Ota, K. 1992. Calcium, magnesium and aluminum concentrations in Parkinson's disease.*Neurotoxicology* 13: 593 – 600.

Yuan, C.Y., Hsu, G.S.W. and Lee, Y.J. 2011. "Aluminum alters NMDA receptor 1A and 2A expression on neonatal hippocampal neurons in rats." *Journal of Biomedical Sciences* 18: 81.

Yuan, C.Y., Lee, Y.J. and Hsu, G.S.W. 2012. "Aluminum overload increases oxidative stress in fore functional brain areas of neonatal rats." *Journal of Biomedical Sciences* 19: 51.

Yumoto, S., Nagai, H., Matsuzaki, H., Matsumura, H., Tada, W., Nagatsuma, M. and Kobayashi, K. 2001. Aluminium incorporation into the brain of rat fetuses and sucklings. *Brain Research Bulletin* 55: 229 – 234.

Zatta, P., Ibn-Lkhayat-Idrissi, M., Zambenedetti, P., Kilyen, M. and Kiss, T. 2002. In vivo and in vitro effects of aluminum on the activity of mouse brain acetylcholinesterase. *Brain Research Bulletin* 59: 41 – 45.

In: Aluminum Neurotoxicity
Editor: Ankica Jelenković

ISBN: 978-1-63484-739-1

Chapter 4

The Protective and Dose-Dependent Effects of L-NAME in Aluminium-Induced Neurotoxicity

Ankica Jelenković[1,*], Marina D. Jovanović[2], Nataša Petronijević[3] and Toplica Lepić[4]

[1]University of Belgrade, Institute for Biological Research "Siniša Stanković", Belgrade, Republic of Serbia
[2]Military Medical Academy, University of Defense, Institute for Medical Research, Belgrade, Republic of Serbia
[3]University of Belgrade, School of Medicine, Institute of Biochemistry, Belgrade, Republic of Serbia
[4]Military Medical Academy, University of Defense, Clinic for Neurology, Belgrade, Republic of Serbia

Abstract

Numerous research studies have undoubtedly shown that aluminium is a very harmful substance which enters the human body externally from the environment. The aluminum intake usually happens unintentionally, due to the fact that people know little about its prevalence in water, factory-processed foods, medicines, cosmetics, etc. When accumulated in human organs, it can cause severe damage, and even lead to chronic neurodegenerative diseases, including Alzheimer's disease. The extent to which nitric oxide (NO) is involved in the basic mechanisms of aluminum neurotoxicity, like oxidative stress and the antioxidant defense, is intriguing scientific community. In this chapter are presented results of the intrahippocampal application of aluminum chloride and a pretreatment with Nω-nitro-L-arginine methyl ester (L-NAME), the non-selective inhibitor of nitric oxide synthase activities. It turned out that, in order to avoid erroneous conclusions about NO not being involved in aluminium-induced neurotoxicity, it was

* Corresponding author: Ankica Jelenković, Institute for Biological Research "Sinisa Stankovic". Bulevar despota Stefana, 142, University of Belgrade, 11000 Belgrade, Republic of Serbia. Phone/fax: +381 11 3420606; E-mail: jelaka@yahoo.com.

necessary to titrate the dose of L-NAME (1, 10 and 100 micrograms). Among the three doses applied prior to the application of aluminum, only the highest dose acted as an antioxidant in the four examined brain structures.

Keywords: aluminium neurotoxicity, nitric oxide, oxidative stress, neurodegenerative diseases

ABBREVIATIONS

NFT	neurofibrillary tangles
ATPase	adenosine triphosphatase
NOS	nitric oxide synthase
FbC	forebrain cortex
S	striatum
BFb	basal forebrain
H	hippocampus

INTRODUCTION

Aluminium (Al) is a well-documented toxin in humans. Known, as well as unknown exposure, usually from the environment, poses a general public health hazard. But, the pathological concentration of aluminium cannot be accumulated in humans (in bones, spleen, kidneys, liver, brain) as long as there is a sufficient elimination, which is mediated by kidneys. However, once accumulated, aluminium can take years to be eliminated because of its very long elimination half-life and storage in more than one compartment [1]. Thus, individuals with altered renal function can be more susceptible to aluminium toxicity which makes a low aluminium diet obligatory for them.

Increasing evidence pointed out to aluminium accumulation as a cause of dysfunction in many tissues. Although this is a controversial topic related to neurodegenerative diseases, the neurotoxicological potential of aluminium exposure was shown not only in animals, but also in humans [2]. A wide array of studies have been conducted to determine aluminium impact on brain aging, and the etiology/pathogenesis of chronic neurodegenerative diseases, such as Alzheimer's (AD), Parkinson's (PD) and Huntington's disease (HD), multiple sclerosis (MS), etc., which are likely to be caused by a mix of genetic, environmental, and lifestyle factors. However, the association between the accumulated aluminium and neurodegenerative diseases today is strongly proposed [3-7]. Although some progress has been made in this regard, the causality relation needs further research, as well as an unequivocal recognition of aluminium hazards.

Originally described by Alois Alzheimer in 1907, AD has emerged as the most common type of dementia in the elderly today. The hypothesis of the potential role of aluminum in AD dates back to 1965 [8]. That could be attributed to the pathological features of aluminium toxicity which can be the same or which can only resemble the changes characteristic for AD, such as aggregation of a amyloid-β peptide into the senile plaques placed extracellularly or in the walls of cerebral blood vessels, and the generation of neurofibrillary tangles in the neurons and glial cells. Even though considerable data concerning this matter exist, the evidence that aluminium causes brain pathology distinctive for AD is still inconclusive and a matter of scientific interest [7].

The neurotoxicity of aluminum could be manifested in the form of biochemical, structural, and neurobehavioral changes in the brain. There are multiple cellular levels for that due to multifaceted targets of aluminium and the facts that it readily binds to different biological macromolecules. Aluminium could "influence more than 200 biologically important reactions" [9]. The involvement of aluminium in oxidative stress, neuroinflammation, impaired membrane functions, competitive displacement of some biomolecular cations from their binding sites extra- and intracellularly (like calcium, iron and magnesium), the dysfunction in neurotransmission and cell signaling, the damage of DNA and other proteins, etc., has been registered so far [10-12].

Nitric oxide (NO) is a gaseous molecule produced in the brain and in other parts of the body under the influence of different stimuli It was named the molecule of the year in 1992 by the *Science* journal [13]. There is no doubt that NO is involved in the multiple processes of health and diseases. Whether NO is involved in the pathogenesis of aluminum neurotoxicity is the subject matter of many research studies. Whether NO is involved in the pathogenesis of aluminum neurotoxicity is a subject matter of many research and especially when it is consider the involvement of nitric oxide signaling pathway in the processes of learning and memory, which were proposed for the first time in 1997 [14]. A number of cognitive processes and functions are damaged in aging and neurodegenerative diseases [15-16]. Similar cognitive deficit were found in animals as the result of aluminium neurotoxicity [17-18]. Because of that, the inhibition of NO synthesis in animal models for various diseases, including transgenic animals (knockout rodents for various nitric oxide synthase isoenzymes), provide a basis for better understanding of NO involved in the health maintenance and the generation/development of neurodegenerative diseases, including aluminium impact on brain functions. These models also provide the ability to investigate the effects of some substances/products in the reversion of various deteriorated biochemical and clinical disturbances in the brain, among which are the impaired cognitive functions induced by aluminium [18-20]. This is important since the insufficient efficiency of products that are used in neurodegenerative diseases today, which provide some symptomatic delay of the disease, but never arrest its progression or lead to remission.

The focus of this chapter will be on the biochemical changes in certain brain structures of rats which were intrahippocampally treated with aluminium chloride. These changes concern the lipid peroxidation and reduced glutathione, as reliable indicators of oxidative stress and defense, respectively. Furthermore, it will be focused on the issue whether the reduction of NO synthesis could influence the changes caused by aluminium, as well as whether this influence is dose-dependent.

NITRIC OXIDE

Nitric oxide is a small, membrane-permeable gaseous molecule synthesized within the cell which can diffuse widely into the extracellular space (~400 μm). Evidence has been provided concerning its complex and multidimensional effects in mammals that are due to its ubiquitous involvement in the inter- and intracellular signalization both within vital physiological, as well as pathological events. All this suggest its double-edged roles and subtle balance between protective and toxic properties. These dual and diametrically opposed effects of NO depend on many factors and conditions, among which there could be the existing NO levels [21].

Nitric Oxide Synthesis

The activity of NO depends on dynamic regulation of its biosynthetic enzyme nitric oxide synthase (NOS). NO is produced from its precursor L-arginine. NOS catalyses reaction of five-electron oxidation of the guanidine group of L-arginine and gives L-citruline and nitric oxide (NO•) as the final products. In this reaction oxygen and nicotinamide adenine dinucleotide phosphate (NADPH) are consumed.

Nitric oxide synthases (EC 1.14.13.39) are a family of enzymes. There are three principal isoforms of NOS in mammals, which show structural homological sequences up to 60% [22]. They are encoded by genes located on different chromosomes. NOS isoforms are usually described by their tissue-specific expression patterns but such distribution can not be applied strictly. It means that many tissues express one or more of NOS isoforms, whose activity is physiologically regulated by a number of compounds and processes within cells [23]. There are:

- neuronal NOS (nNOS, NOS-I, NOS-1, predominantly located in neurons, although it is found in glial cells, i.e., microglia and astrocites, and in some blood vessels as well),
- inducible NOS (iNOS, NOS-II, NOS-2, produced as the response to inflammatory stimuli which is formed in activated macrophages/dendritic cells, immune cells; in the brain it is formed in microglia that can function as macrophages, in oligodendrocytes and astrocytes),
- endothelial NOS (eNOS, NOS-III, NOS-3, located in vascular endothelium),
- mitochondrial NOS (mNOS) is stiil controversial [24]. Its existence was suggested in 1997 for the first time [25]. Its different isoforms have been described in rat and mouse tissues (liver, skeletal muscle, heart, brain, etc). Mitochondria have been demonstrated to be a source of NO as the result of stimulated mNOS by mitochondrial calcium, and also a place of peroxynitrite ($ONOO^-$) production [26].

The activity of NOS enzymes can be inhibited by the application of nitro analogues of L-arginine, such as nitro-L-arginine methyl ester; nitro-L-arginine and monomethyl-L-argjnine, and other compounds [24]. It has been implicated NO modulators to be effectively used as therapeutic agents [27-29].

The both, nNOS and eNOS, are constitutively expressed. They are calcium-dependent enzymes, i.e., they need elevated cytosolic level of calcium that interacts with calmodulin, a binding motif of two domains of NOS [30]. The consequence of making complex calcium-calmodulin is NOS activation with NO transiently synthesized and released. An increased and decreased concentration of calcium switches the cNOS activity on and off.

The release of NO by constitutive NOS (cNOS) occurs as a response to stimulation of several membrane bound receptors, among which postsynaptic N-methyl-D-aspartate (NMDA) subtype of glutamate receptors takes a prominent place [31]. Its stimulation by glutamate, an excitatory amino acid, leads to opening calcium channels. Its transient influx in the cell triggers intracellular signaling pathways and activation of calcium-sensitive enzymes, including, but not limiting, the activation of calcium-calmodulin-dependent enzymes, such as nNOS and eNOS.

On the other hand, iNOS expression is enhanced and synthesized upon inflammation and is calcium-independent, i.e., it is activated at the basal calcium concentration and generates large amounts of NO (micromolar range) for prolonged period of time depending on the noxious factor that triggered it. However, recent research has shown inconsistencies of constitutive and inducible NOS classification [32].

The histochemical marker of nitrergic neurons was found to be a nicotinamide adenine dinucleotide phosphate (NADPH)-diaphorase, indicating co-localization of the nNOS with NADPH-diaphorase, as well as the neuroanatomical distribution pattern of nNOS and nitrergic neurons [33]. In spite the fact that the nNOS positive neurons constitute only 1 to 2 percent of cells in the brain regions, at the most localized in such as the cerebral cortex, the hippocampus and striatum, they have extensive influence on the majority of neurons in the brain.

Nitric Oxide Actions

NO• is a free radical whose direct toxicity is considered to be modest. Tissue oxygen rapidly oxidizes it into stable end products like nitrates and nitrites. Further on in this text, NO will be used in the meaning of NO as a free radical.

NO and NO-derived species have diverse biological effects such as the inhibition/ activation of enzymes, receptors, transporters and membrane channels. These species are also involved in the protein aggregation, impairment of cellular signaling, mitochondrial dysfunction, DNA and membrane injury, etc. [34-35]. This could be partly attributed to the ability of NO to:

- stabilize its unpaired electron:
  - by reacting with free radicals or
  - by binding to certain transition metal ions in various molecules of metalloproteins as potential targets, and thus forming complexes with them. These metal-containing proteins can be either stimulated (e.g., guanylate cyclase) or inhibited (eg., cytochrome c oxidase, complex IV). These interactions could be reversible, but also toxic, as in the case of aconitase, for example.

A well-established mechanism by which NO manifests its effects is its reversible interaction with the heme moiety of soluble guanylate (guanylyl) cyclase (sGC). It is the major, primary and the most sensitive physiological target ("receptor") for NO in the NO-cGMP signaling pathway. This interaction is achieved at very low concentrations of NO (picomolar range) that typically exist in the unstressed cell when it does not readily form damaging species [36]. NO promotes sCG activation that mediates the production of cGMP and increase its concentrations, thus leading to cGMP-dependent signaling cascade, which is involved in numerous physiological functions, including synaptic plasticity and smooth muscle relaxation.

- NO, by itself, can directly modify a number of proteins (protein nitrosylation), including NMDA receptors, nucleic acids, etc. S-nitrosylation formed under the basal low level of NO is protective, whereas that induced upon stressed conditions is implicated to be harmful [29, 37].
- NO can modify redox-sensitive thiols on target enzymes, which then change their function and structure. This is particularly important from the standpoint of neurodegenerative diseases [38].

The reversibility of these reactions depends on the duration of the exposure to NO and on its concentration. Irreversible reaction dominates and could be toxic when NO is overproduced, or when the antioxidant defense is insufficient. While the reaction with NO radical is named nitrosylation (R-NO•), nitrosation is the reaction with NO (R-NO), and nitration is the reaction with nitro group (R-NO_2) [39].

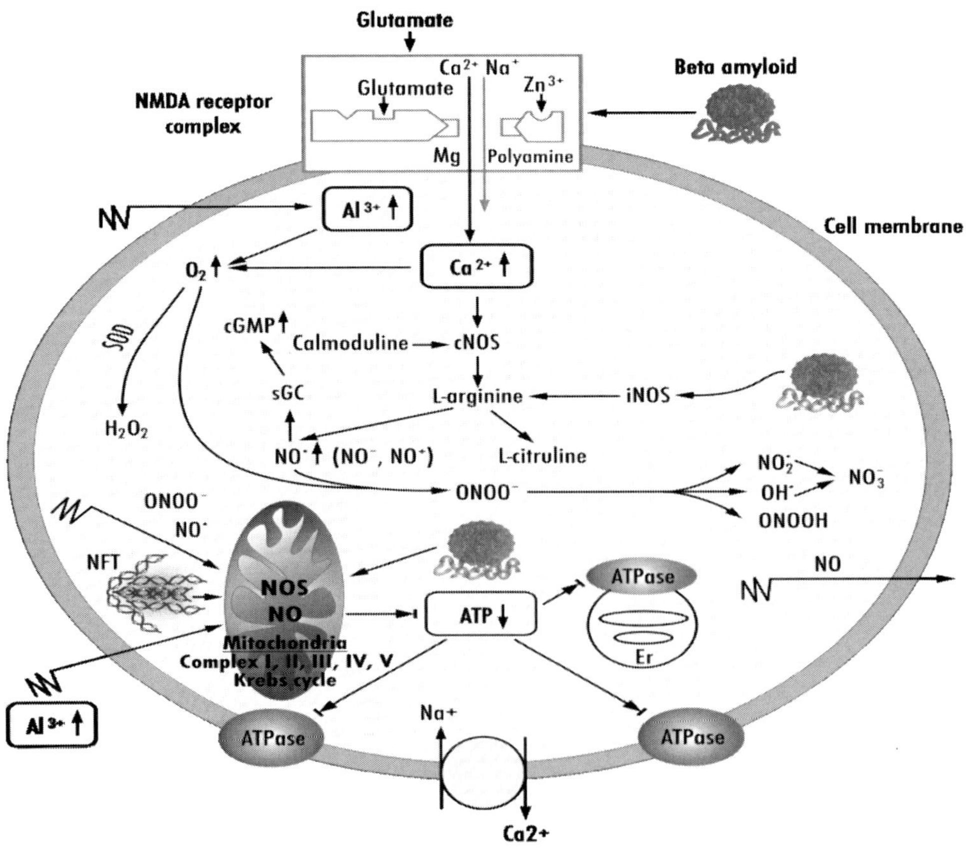

Figure 1. The influence of NO and aluminium in the generation/development of neurodegenerative diseases, like Alzheimer's disease (the meaning of the abbreviations exists in the text).

Nitric Oxide: Mitochondria and Glycolysis

Nitric oxide is a potent modulator of mitochondrial functions, primarily of the respiratory chain and energy (adenosine triphosphate - ATP) synthesis (Figure 1) [40]. NO reacts with metals (iron and copper ions) at the catalytic active sites in the functional groups of the mitochondrial complexes. These effects are not only caused by the NO activity, but also by the NO-derived secondary species, such as $ONOO^-$ and S-nitrosothiols (RSNO).

- NO interacts with complex IV, which is, among the five, the most sensitive component of the respiratory chain to it. NO competes with oxygen for the ferrous ion, thus inhibiting complex IV. This reaction is reversible in physiological conditions, when NO concentration is low, and when it causes and activates cellular signaling cascades. The interaction of NO with complex IV regulates the oxygen uptake and the production of superoxide anion from the mitochondrial electron transfer chain, and also the production of ATP, which is generated *via* oxidative phosphorylation. This biological action of NO leads to the reversible inhibition of ATP synthesis and energy conservation [41-43];
- NO reversibly inhibits complex I and III of the respiratory chain;
- The mitochondrial matrix contains higher levels of manganese superoxide dismutase (Mn-SOD) than the other parts of the cell. It removes the superoxide radical, which is generated in the mitochondrial matrix. However, $ONOO^-$ can inhibit Mn-SOD activity (nitrated Mn-SOD) which, then, leads to mitochondrial injury. The increased levels of the last compound were found in the brain of AD, PD, and patients with amyotrophic lateral sclerosis (ALS) [44];
- NO regulates glycolysis *via* cGMP-independent mechanisms. It is enhanced in astrocytes when the mitochondrial respiration is inhibited. In that way NO prevents ATP depletion;
- NO protects the mitochondrial membrane from the damage during excitotoxicity, when the excessive calcium influx into mitochondria arises, by which their membrane integrity is compromised with consequential cell damage and death. Thus, NO is involved in cell survival and cell death;
- NO, *via* $ONOO^-$, stimulates the activity of glucose-6-phosphate dexydrogenase (G6PDH), the rate-limiting enzyme in the pentose-phosphate pathway, which leads to the regeneration of NADPH, the principal molecule in the reductive reactions in the cell, including the reduction of glutathione from its oxidized form [42];
- $ONOO^-$ can irreversibly inactivate the electron transport chain complex I, II, and V (ATP synthetase) being harmful for oxidative phosphorylation;
- $ONOO^-$ impairs the Krebs cycle. It blocks aconitase activity [45].

The effects of NO and $ONOO^-$ on glycolysis, G6PDH and mitochondrial complexes of respiratory chain make them the neuroprotective substances that could be harmful under pathological conditions [42-43].

The Reactive Nitrogen Species and Nitrosative Stress

The intracellular redox status is determined by the balance between the oxidant and antioxidant system. Similar to the reactive oxygen species (ROS), which includes superoxide anion ($O_2^{\bullet -}$) and a family of superoxide anion-related molecules, the scientific community has coined the phrase reactive nitrogen species (RNS). The RNS group includes compounds generated from the oxidation or the reduction of NO, like: nitric oxide free radical ($NO^{\bullet}$) and $NO^{\bullet}$-derived congeners such as peroxynitrite anion ($ONOO^-$), dinitrogen trioxide (N_2O_3), nitrogen dioxide ($NO^{\bullet}_2$), nitroxyl anion (NO^-), nitrosonium cation (NO^+), nitrite (NO_2^-), nitrate (NO_3^-), etc. [39].

The overproduction of RNS or the deregulated formation of NO and NO-derived RNS is called nitrosative stress, which is typically seen in the stressed intracellular environment [47 Hausladen]. It is well known that ROS and RNS play a dual role since they can be both beneficial and deleterious species, depending on the levels of their production and removal [46, 34]. As they can lead to oxidative and nitrosative stress (ONS), the both ROS and RNS are involved in the pathogenesis of neurodegenerative diseases, such as AD, PD, ALS, HD [35, 48].

The basal NO levels are too low to effectively compete for the superoxide anion with Superoxide dismutase (SOD), which converts superoxide to the less damaging hydrogen peroxide and diminish opportunity to NO for the forming RNS. However, although not being well understood yet, NO easily reacts with superoxide anion without enzymes and forms $ONOO^-$ anion in the case of excessive NO generation, when the rate of superoxide reacting with NO is several times faster than the one with SOD. Such NO generation could be developed as the result of the overstimulation of postsynaptic NMDA receptors, which consequently activates constitutive NOS, or it could be the result of the overstimulation of the calcium independent iNOS. All of the above mentioned conditions are registered in the variety of disturbances, such as neuroinflammation, excitotoxicity, aging and age-related neurodegenerative diseases. In line with this are the data about the increased NOS activity being registered in these diseases [22, 49].

$ONOO^-$ is not a free radical but an extremely reactive molecule which is a powerful oxidant. $ONOO^-$ is thought to be responsible for the NO-mediated neurotoxicity as a great number of biomolecules can be oxidized and/or nitrated by $ONOO^-$ and $ONOO^-$-derived radicals. As a very unstable molecule, at the physiological pH $ONOO^-$ rapidly becomes peroxinitrous acid (ONOOH), which is decomposed into a highly toxic hydroxyl-like radical and NO_2. In spite of NO (~1 s) and $ONOO^-$ (~10 ms) being estimated to have short half-lives, which vary under the influence of many factors, they are still sufficiently long for their effects to manifest.

According to the findings of numerous studies, nitrosative stress and protein nitration modify structure and, thereby, the activity and normal function of certain proteins can be compromise. That can consequently lead to protein aggregation. Such forms of proteins are very toxic and could provoke neurodegeneration [50].

NITRIC OXIDE IN THE BRAIN

In recent studies, a new light has been cast on the role of NO and NO-derived species in a variety of physiological processes in the brain, which radically improved the knowledge and understanding of neural communication [51]. NO belongs to non-classical neurotransmitters. What this means is that:

- NO is not stored in synaptic vesicles, but it depends on new synthesis and is synthesized enzymatically on demand;
- it is not released by exocytosis after the membrane depolarisation, but it diffuses through the cell membrane, thus influencing neural and non-neural cells in the vicinity of the cell producing it;
- it does not act upon the typical extracellular receptor proteins on synaptic membrane, but it has a direct effect on some targets both within and out of the cell (heme group, for example);
- it is not reuptaked or metabolized like conventional transmitters in the process of the termination of their activity, but its activity is terminated after its chemical reaction with other molecules, which then leads to the creation of other compounds. This means that the activity of NO is primarily controlled by its synthesis.

In physiological conditions:

- NO can activate a number of signaling pathways as a messenger with the neurotransmission-like function;
- NO acts pre-sinaptically (retrograde activity) and can participate in the regulation of neurotransmission (dopamine, noradrenaline, serotonin, acetylcholine, glutamate, gamma-aminobuturic acid: GABA, etc.) [11, 45];
- NO can be involved in the development of synaptic plasticity:
  - NO facilitates the long-term potentiation (LPT). It is an accepted molecular mechanism of learning and memory where NO, acting as a retrograde messenger, induces LTP. This is the result of the presynaptic NO-enhancend neurotransmitter release and potentiating postsynaptic response [52],
  - NO may also be involved in the long-term depression (LTD) with the reduction in the synaptic strength;
- NO is involved in the prevention of glutamate excitotoxicity, a state of extensive release of glutamate into the synaptic cleft or the overstimulation of glutamate receptors located postsinaptically. Due to the retrograde activity of NO and its interaction with NMDA receptors, the activity of these receptors is decreased, i.e., the down regulation of NMDA receptors and NO synthesis occurs. This bioactivity feedback of NO is neuroprotective;
- NO enables the optimal cerebral blood flow through the modulation of the vascular tone;
- NO is involved in immune responses, etc.

The modulation of nitric oxide levels exerts harmful effects in aging and neurodegenerative diseases. Its exact mechanism is not fully understood, but weak and sustained excitotoxicity has been proposed to occur that leads to permanent different damages in the brain. Data about the involvement of NO in brain pathologies are related to [53-54]:

- Aging
- Alzheimer's disease
- Parkinson's disease
- Huntington's disease
- Amyotrophic lateral sclerosis
- Multiple sclerosis
- Depression
- Deterioration of cognitive functions, etc.

Different brain cells are not equally vulnerable to NO and $ONOO^-$ -mediated toxicity. This is due to their dissimilar antioxidant abilities, which are especially related to the status of glutathione and other antioxidant enzymes (SOD, GSH peroxidase, catalase), and that is why the glial cells are more resistant to NO and NO-derived toxicity than neurons.

THE GLUTATHIONE SYSTEM

Thiols, also called sulfhydryls, represent major compounds in the regulation of the cellular redox homeostasis. Out of four thiol-containing amino acids: cysteine, methionine, homocysteine, and taurine, glutathione (GSH, the abbreviation usually used to designate reduced form of glutathione), is a cysteine-containing antioxidant, which also includes the glutamate and glycine in its molecule. It was discovered in yeast in 1888, and synthetized for the first time in 1935 [44].

This most abundant nonprotein thiol in cells is *de novo* synthesis in cytosol and transported into the mitochondria, nucleus, endoplasmatic reticulum, etc. and into the intracellular spaces, as well. Glutathione is critical for cellular antioxidant defense since its intracellular concentration is dominant and is 100 -1000 times higher than any other redox-modulatory system. With its extremely high redox potential, GSH is a part of overall cellular antioxidant defense system, as well as a part of apoptosis and cell survival [55]. Sulfhydryl residues in GSH are able to be easily oxidized to form disulfide (GSSG), which is an oxidized form of glutathione. Thus, under oxidative stress GSH can reversibly form disulfide bonds with protein thiols (S-glutathionylation) to prevent protein oxidation.

In the brain, GSH is synthesized in the astrosytes and neurons, whereby higher levels are reported to be found in astrocytes. The dysregulation of GSH redox system may contribute to the development of neurodegenerative diseases [56].

The effects of GSH are the result of several mechanisms:

- The high antioxidant potential
  - GSH is a free radical scavenger in the detoxification of the ROS and RNS. It acts as a potent antioxidant against a variety of reactive oxidative and nitrosative

species, as well as against the products of ROS-induced lipid peroxidation (malondyaldehyde, for example). GSH directly, nonenzimatically, reacts with superoxide, NO, $ONOO^-$, and hydroxyl radical, resulting in their removal.
  - Considering high intracellular concentrations of S-nitrosoglutathione (GSNO), its protective effects, resulting in the reducing levels of $ONOO^-$, are to be expected. Furthermore, by reducing the levels of $ONOO^-$ in pathological conditions, GSNO decreases neuronal cell death. Also, "the GSNO could be a reservoir of the endogenous signaling molecule NO which maintains the physiological reactions that deserve the involvement of NO" that it is generated under low (submicromolar) concentration of NO [37, 57].
- The metabolic regulation
  - GSH is an ubiquitous cofactor for the glutathione dependent enzymes, (a family of peroxidases, S-transferases, thioltransferases, etc.), where it acts as a donor of reducing equivalents [56].
- The influence on the glutamatergic transmission
  The oxidized and reduced form of the physiological concentration of glutathione directly and indirectly modulates the glutamatergic ionotropic receptors: NMDA, AMPA (2-amino-3-hydrohy-5-methyl-4-isoxazoleprpionate) and kainate. In this way, glutathione acts neuroprotectively against exitotoxicity [58].
- The detoxification of xenobiotics.

The reduced form of glutathione is enzymatically regenerated from GSSG by glutathione reductase. The cytosolic ratio between GSH and GSSG (GSH:GSSG) in the steady state is typically higher than 100:1, which makes GSH a major antioxidant. However, the redox state in the endoplasmatic reticulum, where proteins are being folded, is more oxidative, since GSH:GSSG ratio is about 5:1. This is important for the development of neurodegenerative diseases, in which misfolding proteins are one of the hallmarks [59].

The decreased GSH:GSSG ratio is observed in the oxidative stress. In such conditions, it could be in cytosol even less than 10. Such large oscillations are a reliable indicator of its great capability to maintain redox homeostasis. Thus, GSH depletion could be caused by the increased oxidative stress and/or the decreased GSH synthesis/regeneration, especially in the neurons, which are more vulnerable to free radicals than glial cells due to multiple reasons. The deficiency of GSH contributes to the appearance of oxidative/nitrosative stress and its deterioration. The decline of GSH is found in neurological diseases, including AD, in which its depletion precedes neurodegeneration [44].

The Protective and Dose Dependent Effects of L-NAME in Aluminium-Induced Neurotoxicity

Material and Methods

Animals and Experimental Design

Male adult Wistar rats, 12 weeks old at the beginning of the experimental protocol, were used for this study. The animals were allowed access to commercial rat food and tap water *ad libitum,* and were maintained in controlled environmental conditions: it was obtained 12 h/12 h light/dark cycle, room temperature 23 ± 2°C. The research protocol was approved by the local ethics committee for animal experimentation (Military Medical Academy, Belgrade, Serbia), which is in harmony with international regulations.

After being intraperitoneally anesthetised with sodium thiopental (Specia, Paris) (0.04 g/kg body weight), rats were placed on a stereotaxic instrument for small animals for the hippocampal administration of the examined substances. The solutions of the examined substances were injected by Hamilton micro syringe into the left brain hemisphere Cornu Ammonis region 1 (CA1) of the hippocampus, a limbic system structure. The position for the injection was determined relative to the lambda suture, defined from its centre: 3.1mm dorsally, 4.3 mm laterally and 2.5 mm ventrally from the skull surface [60]. The recovery period after the treatments lasted for 12 days. Thereafter the rats were sacrificed by decapitation. The heads were flash-frozen in liquid nitrogen and stored at -70°C until brain samples were prepared for biochemical analysis.

Treatments

Six groups of rats, each containing 9-12 animals, were used. They were randomly subjected to treatments. The first group received 0,9 per cent w/v NaCl (NaCl, control group) and the second one aluminium chloride (Sigma-Aldrich, USA) at 3.7×10^{-4} g/kg body weight dissolved in sterile deionized water (AlCl3 group). Before aluminium chloride administration, the rats in the third, fourth and fifth group were treated with Nω-nitro-L-arginine methyl ester (L-NAME), in the doses of 1, 10 and 100 micrograms, dissolved in saline solution. The animals in the sixth group received 100 micrograms of L-NAME alone. All the examined substances were applied in the volume of 0.01 mL.

Biochemical Analyses

Biochemical analyses were performed in crude mitochondrial fractions from the ipsilateral brain regions to the injection site that were kept on ice while being dissected [61]. The index of lipid peroxidation [62] and the level of reduced glutathione [63] were determined in four brain regions: the forebrain cortex (FbC), striatum (S), basal forebrain (BFb) and hippocampus (H).

Statistical Analysis

Data were expressed as means ± SD. The Student's t-test was used for comparisons between groups. Statistical significance was determined at $p < 0.05$.

Results and Discussion

In all of the four examined brain structures, when aluminium was applied and NO synthesis was decreased by L-NAME, both the inevitable detrimental effects of aluminium and the protective effects against aluminium toxicity, were observed respectively. Such extensive spatial spreading of the effects of the applied substances in the CA1 sector of the hippocampus was the result of very complex anatomical and neurochemical communications within the brain.

The Neurotoxic Effects of Aluminium

The disruption of essential metal homeostasis, oxidative stress and oxidative protein modification are documented in neurological disorders associated with aluminium overload [64-65, 18]. In fact, the most thoroughly examined and documented are the pro-oxidant effects of aluminium, together with neuroinflammation. The lipid peroxidation is one of the consequences of aluminium-induced free radical overproduction, as well as the decreased activity of antioxidant enzymes. It results in the dysfunction and damage of biological membranes. This is preceded by the attempt of the antioxidant mechanisms to suppress the development of oxidative stress, which then leads to their consumption, one of which is GSH.

The deleterious effects of aluminum were present at the application site (hippocampus), as well as in the other three structures: the forebrain cortex, striatum and basal forebrain. The index of lipid peroxidation, an indirect and reliable marker of oxidative damage, was significantly increased in all the examined brain regions in comparison to control group ($p < 0.01$ in the forebrain cortex, $p < 0.05$ in other three structures (Figure 2). In spite of the fact that the index of lipid peroxidation does not allow us to exactly determine the source of reactive oxygen and nitrogen species, as well as the exact time when this damage was developed, in some studies the increased index of lipid peroxidation was registered very early after aluminium administration. It occurred as early as three hours after the treatment, and lasted up to 30 days after the acute application of aluminium into the hippocampus, indicating its long lasting deleterious effects [66-67].

By affecting the iron homeostasis, aluminium possesses detrimental effects on enzymes of the electron transport chain of oxidative phosphorylation, like complexes I, III and IV, as well as on some enzymes of the tricaboxcyclic acid cycle [18, 12]. Hence, it could increase the production of free radicals, deteriorate the production of a cellular energy molecule (ATP) and reductive equivalents.

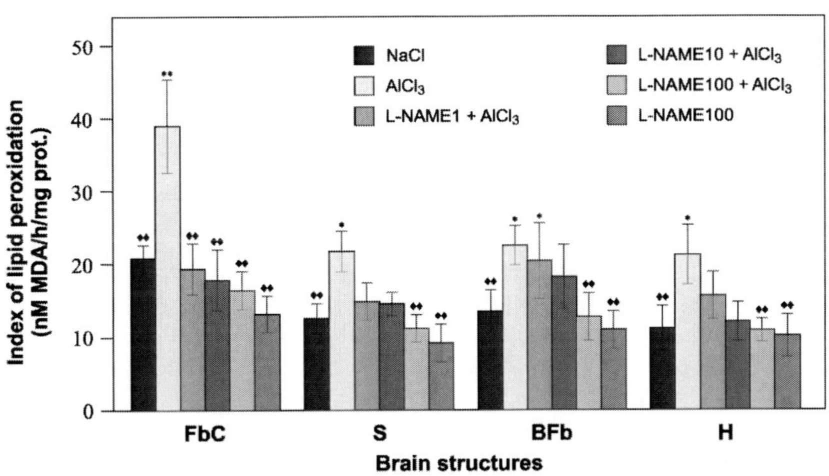

Figure 2. Index of lipid peroxidation in the brain of Wistar rats (n = 9 – 12) intra-hippocampally treated with 0.9% NaCl, aluminium chloride ($AlCl_3$) and L-NAME 1, 10 and 100 micrograms administered prior to AlC13.

The values were expressed as mean ± SD. $p < 0.05$, $p < 0.01$ are the levels of statistical significance found versus 0.9% NaCl (*, **) and versus $AlCl_3$-treated rats (♦, ♦♦) (Student's t-test). The rats were sacrificed 12 days after treatments.

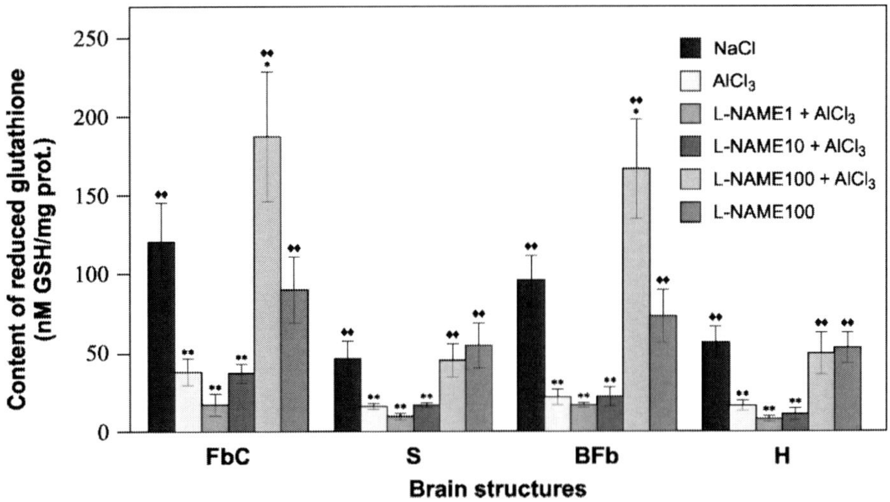

Figure 3. Content of reduced glutathione in the brain of Wistar rats (n = 9 – 12) intra-hippocampally treated with 0.9% NaCl, aluminium chloride ($AlCl_3$) and L-NAME 1, 10 and 100 micrograms administered prior to AlCl3.
The values were expressed as mean ± SD. $p < 0.05$, $p < 0.01$ are the levels of statistical significance found versus 0.9% NaCl (*, **) and versus $AlCl_3$-treated rats (♦, ♦♦) (Student's t-test). The rats were sacrificed 12 days after treatments.

A very reliable indicator of the presence of oxidative stress in aluminium neurotoxicity was the actual finding of the extremely high decrement of reduced glutathione content in all the observed structures ($p < 0.01$ in all structures, compared to the control group, Figure 3). The neuronal vulnerability to free radical damage is dictated by GSH, the most important defense mechanism in the body, and it is the primary protector of mitochondrial membranes against oxidative damage. The changes of GSH levels over time, starting from three hours after the application of aluminum, up to 30 days after the application of this neurotoxin, indicate its important role in aluminium neutoxicity and its consumption over time [66].

All brain structures are not simultaneously affected by oxidative stress, but there is a specific sequence in that. The hippocampus always appears to be an early target of the age-related and AD changes of biochemical and structural neuropathology. Moreover, it was also shown that the per-orally applied aluminium chloride developed oxidative stress in the hippocampal mitochondria at first. These damages precede the similar disturbances in other brain structures [68]. This clearly points to the hierarchy of the brain structures susceptibility to aluminum toxicity, which is similar to that seen in AD. In addition to other findings, hippocampus, the primary site of short-term memory initiation and the first step for memory consolidation, is among the earliest brain structures affected in both cases [68-69]. All the aforementioned changes in the brain induced by aluminium, support the therapeutic approach with antioxidants.

The Effects of L-NAME in Healthy and Aluminium Treated Rats

L-NAME did not demonstrate any effects when applied alone (in healthy, not aluminium-pretreated rats), neither in relation to the index lipid peroxidation nor to GSH level. On the other hand, L-NAME demonstrated strong effects related to both parameters when it was applied before aluminium.

There are many mechanisms through which aluminum causes damaging effects. One of them is the modulation of NOS activity and NO production, with changes of nitrite levels over time, where nitrites represent a measurable indicator of generated NO [70]. In line with this was the fact that the production of NO obviously increased very quickly, within the period of 10 minutes after the aluminium application [65]. However, a statistically significant increment was registered on the third day after the treatment. The increase of NO production could be partly the result of the increased intracellular levels of calcium, since aluminium strongly affects calcium influx and efflux and, thus, disrupts calcium homeostasis [10]. In these conditions, there was a possibility of an increased NOS activity together with excessive NO production, which were likely to provoke the mitochondrial dysfunction. Furthermore, the increased iNOS staining was found in aluminium intoxication indicating that aluminum causes harmful effects by activating iNOS as well [71]. All of the aforementioned was in harmony with the findings of a stronger inflammatory reaction in the hippocampus in the aluminium-treated animals, when compared to the controls [65].

Nitrite levels, however, were not increased later on, that is, on the 12th and 30th day after the aluminium treatment [65-67, 72]. This could be the result of a successful antioxidant mechanism defense, in spite of the long lasting inflammatory reaction [65]. The reliable indicator of antioxidants spending and/or disrupted regeneration was the almost depleted (GSH) in all of the observed structures (Figure 3). This occurrence could be in line with the facts documenting the interaction between NO and $ONOO^-$ with GSH and the generation of S-nitrosoglutsathione, since both of them, NO and $ONOO^-$, strongly oxidize sulphydryls, including GSH, representing the way in which the cell fights against the damaging effects of NO and $ONOO^-$.

The increased production of NO, as an inseparable part of the pathogenesis of aluminum neurotoxicity, was fully demonstrated with the application of L-NAME, the non-selective NOS inhibitor, which was applied before aluminum. This becomes evident only when examining multiple doses of L-NAME. Otherwise, the conclusion concerning the non-involvement of NO in aluminium neurotoxicity could be quite correct when taking into account the effects of the two lower doses of L-NAME (1 and 10 micrograms). When compared to the aluminium treatment, both in terms of the impact on lipid peroxidation, as well as on the level of GSH, obtained improvement was no statistically significant. However, when the applied doses of NOS inhibitor "optimally" inhibited the synthesis of NO, it was shown that NO was an integral part of the pro-oxidant effects of aluminium. Indeed, although all of the three doses of L-NAME diminished the aluminium impact on the lipid peroxidation and GSH levels, only the third dose (100 micrograms) led to the statistically significant changes in both of these parameters.

In terms of the index of lipid peroxidation, the highest dose of L-NAME caused its decrease in the all the brain structures ($p < 0.01$ in the forebrain cortex, $p < 0.05$ in other three structures, in comparison to aluminium-treated animals). In the forebrain cortex, two lower doses also decreased the index of lipid peroxidation ($p < 0.01$). The pro-oxidant direction of NO in aluminium neurotoxicity was confirmed by the increment of GSH levels in all the examined structures when L-NAME was applied before aluminium. The increment was very small when two lover doses were applied, while only with the highest it was reached a statistical significance in comparison to the aluminium-treated rats ($p < 0.01$ in all structures). Furthermore, in the forebrain cortex and in the basal forebrain that increment significantly exceeded the values found in the control group of animals ($p < 0.05$). These findings showed a strong participation of GSH when combating the toxic effects of NO in the case of the aluminium intoxication. Although GSNO may represent a reservoir of NO which participates in the signaling reactions, GSH predominantly acted as an antioxidant in our experiment. The opposite direction of the level of the index of lipid peroxidation and levels of GSH in the conditions when L-NAME was administered before aluminium suggests such a role of reduced glutathione. Furthermore, the reversion could also be the result of GSH regeneration from GSSG, since the physiological level of NO, which could be obtained with the inhibition of NO synthesis, activates the pentose-phosphate pathway. This pathway is the most important mechanism for recycling the oxidized NADP to NADPH without which, as a cofactor, the recovery of GSH is not possible.

Aluminium alone, as well as in combination with NO derived from the sodium nitroprusside (SNP, NO donor), showed additional inhibitory effects on mitochondrial aconitase activity and interfered with the metabolic tricarboxylic acid cycle, that is also inhibiting under high concentration of nitric oxide alone [73-76]. So, the NOS inhibition could result in protecting the Krebs cycle from the deleterious effects of NO, which could be beneficial for the antioxidant defense system.

In order to reverse the damaging effects of NO obtained in aluminium toxicity, it was necessary to decrease its levels. It was undoubtedly shown that this reversion could only be possible with the administration of appropriate doses of L-NAME. While all three doses showed almost the same direction against aluminium neurotoxicity in relation to the index of lipid peroxidation and GSH, it was demonstrated that the dose titration was necessary in order to achieve full protective effects of L-NAME.

CONCLUSION

Results reported in this chapter clearly demonstrated that acute intrahipocampal exposure to aluminium chloride evoked oxidative stress in the forebrain cortex, striatum, basal forebrain and hippocampus. Both the index of lipid peroxidation and the level of GSH confirmed its presence on the12^{th} day after the application. These effects were prevented by the inhibition of NO synthesis. That was obtained only if the appropriate dose of the L-NAME, the nonselective NOS inhibitor, was applied. Thus, the pretreatment with such dose of L-NAME clearly demonstrated its protective effect against aluminium neurotoxicity. Otherwise, it might be mistakenly concluded that the reduction of NO synthesis was of no importance and that NO was not involved in this toxicity. The decreased oxidative stress with L-NAME could be the result of an improved mitochondrial function, the principal source of superoxide anion production in health and disease, and also energy metabolism, whose main protector against oxidative damage is glutathione. By clarifying the profiles of nitric oxide in the brain tissues and their participation in physiological and pathophysiological processes we can open a new avenue for the development of new therapeutic strategies.

ACKNOWLEDGMENTS

This research was supported by the Ministry of Science of the Republic of Serbia (contract number 175058) and the Ministry of Defense of the Republic of Serbia (contract number MMA/06-10/B.3).

REFERENCES

[1] Priest, N.D., Newton, D., Day, J.P., Talbot, R.J., Warner, A.J. Human metabolism of aluminium-26 and gallium-67 injected as citrates. *Hum Exp Toxicol.* 1995; 14(3):287-93.

[2] Exley, C. What is the risk of aluminium as a neurotoxin? *Expert Rev Neurother.* 2014; 14(6):589-91.

[3] Frisardi V., Solfrizzi V., Capurso C., Kehoe P.G., Imbimbo B.P., Santamato A., et al. Aluminum in the diet and Alzheimer's disease: from current epidemiology to possible disease-modifying treatment. *J Alzheimers Dis.* 2010; 20:17-30.

[4] Ahmed, S.S., Santosh, W. Metallomic profiling and linkage map analysis of early Parkinson's disease: a new insight to aluminum marker for the possible diagnosis. *PLoS One.* 2010 Jun 22; 5(6):e11252.

[5] Exley, C., Vickers, T. Elevated brain aluminium and early onset Alzheimer's disease in an individual occupationally exposed to aluminium: a case report. *J Med Case Rep.* 2014; 8:41.

[6] Bondy, S.C. Prolonged exposure to low levels of aluminum leads to changes associated with brain aging and neurodegeneration. *Toxicology.* 2014; 315:1-7.

[7] Walton, J.R. Chronic aluminum intake causes Alzheimer's disease: applying Sir Austin Bradford Hill's causality criteria. *J Alzheimers Dis.* 2014; 40:765-838.

[8] Klatzo, I., Wisniewski, H., Streicher, E. Experimental production of neurofibrillary degeneration. *J Neuropath Exp Neurol.* 1965; 24:187-99.

[9] Kawahara, M., Kato-Negishi, M. Link between aluminum and the pathogenesis of Alzheimer's disease: the integration of the aluminum and amyloid cascade hypotheses. *Int J Alzheimers Dis.* 2011 Mar 8; 2011:276393. doi: 10.4061/2011/276393.

[10] Kaur, A., Gill, K.D. Disruption of neuronal calcium homeostasis after chronic aluminium toxicity in rats. *Basic Clin Pharmacol Toxicol.* 2005; 96(2):118-22.

[11] Gonçalves, P.P., Silva, V.S. Does neurotransmission impairment accompany aluminium neurotoxicity? *J Inorg Biochem.* 2007; 101(9):1291-338.

[12] Kumar, V., Gill, K.D. Oxidative stress and mitochondrial dysfunction in aluminium neurotoxicity and its amelioration: a review. *Neurotoxicology.* 2014; 41:154-66.

[13] Culotta, E., Koshland, D.E. Jr. NO news is good news. *Science.* 1992; 258(5090):1862-5.

[14] McCann, S.M. The nitric oxide hypothesis of brain aging. *Exp Gerontol.* 1997; 32(4-5):431-40.

[15] Virarkar, M., Alappat, L., Bradford, P.G., Awad, A.B. L-arginine and nitric oxide in CNS function and neurodegenerative diseases. *Crit Rev Food Sci Nutr.* 2013; 53(11):1157-67.

[16] Chung, K.K., David, K.K. Emerging roles of nitric oxide in neurodegeneration. *Nitric Oxide.* 2010; 22(4):290-5.

[17] Abu-Taweel, G.M., Ajarem, J.S., Ahmad, M. Neurobehavioral toxic effects of perinatal oral exposure to aluminum on the developmental motor reflexes, learning, memory and brain neurotransmitters of mice offspring. *Pharmacol Biochem Behav.* 2012; 101(1):49-56.

[18] Jovanović, M.D., Jelenković, A., Stevanović, I.D., Bokonjić, D., Colić, M., Petronijević, N., et al. Protective effects of glucose-6-phosphate dehydrogenase on neurotoxicity of aluminium applied into the CA1 sector of rat hippocampus. *Indian J Med Res.* 2014; 139:864-72.

[19] Kumar, A., Prakash, A,, Dogra, S. Neuroprotective effect of carvedilol against aluminium induced toxicity: possible behavioral and biochemical alterations in rats. *Pharmacol Rep.* 2011; 63(4):915-23.

[20] Jelenković, A., Jovanović, M.D., Stevanović, I., Petronijević, N., Bokonjić, D., Zivković, J., et al. Influence of the green tea leaf extract on neurotoxicity of aluminium chloride in rats. *Phytother Res.* 2014; 28(1):82-7.

[21] Calabrese, V., Mancuso, C., Calvani, M., Rizzarelli, E., Butterfield, D.A., Stella, A.M. Nitric oxide in the central nervous system: neuroprotection versus neurotoxicity. *Nat Rev Neurosci.* 2007; 8(10):766-75.

[22] Zhou, L., Zhu, D.Y. Neuronal nitric oxide synthase: structure, subcellular localization, regulation, and clinical implications. *Nitric Oxide.* 2009; 20(4):223-30.

[23] Esplugues, J.V. NO as a signalling molecule in the nervous system. *Br J Pharmacol.* 2002; 135(5):1079-95.

[24] Alderton, W.K., Cooper, C.E., Knowles, R.G. Nitric oxide synthases: structure, function and inhibition. *Biochem J.* 2001; 357(Pt 3):593-615.

[25] Ghafourifar, P., Richter, C. Nitric oxide synthase activity in mitochondria. *FEBS Lett.* 1997; 418(3):291-6.

[26] Navarro, A., Boveris, A. Mitochondrial nitric oxide synthase, mitochondrial brain dysfunction in aging, and mitochondria-targeted antioxidants. *Adv Drug Deliv Rev.* 2008; 60(13-14):1534-44.

[27] Whiteley, C.G. Arginine metabolising enzymes as targets against Alzheimers' disease. *Neurochem Int.* 2014; 67:23-31.

[28] Zhang, R.L., Zhang, Z.G., Chopp, M. Targeting nitric oxide in the subacute restorative treatment of ischemic stroke. Ex*pert Opin Investig Drugs.* 2013; 22(7):843-51.

[29] Silverman, R.B. Design of selective neuronal nitric oxide synthase inhibitors for the prevention and treatment of neurodegenerative diseases. *Acc Chem Res.* 2009; 42(3):439-51.

[30] Spratt, D.E., Newman, E., Mosher, J., Ghosh, D.K., Salerno, J.C., Guillemette, J.G. Binding and activation of nitric oxide synthase isozymes by calmodulin EF hand pairs. *FEBS J.* 2006; 273(8):1759-71.

[31] Lourenço, C.F., Ferreira, N.R., Santos, R.M., Lukacova, N., Barbosa, R.M., Laranjinha, J. The pattern of glutamate-induced nitric oxide dynamics in vivo and its correlation with nNOS expression in rat hippocampus, cerebral cortex and striatum. *Brain Res.* 2014; 20:1554:1-11.

[32] Mattila, J.T., Thomas, A.C. Nitric oxide synthase: non-canonical expression patterns. *Front Immunol.* 2014 Oct 9;5:478. doi: 10.3389/fimmu.2014.00478. eCollection 2014.

[33] Hope, B.T., Michael, G.J., Knigge, K.M., Vincent, S.R. Neuronal NADPH diaphorase is a nitric oxide synthase. *Proc Natl Acad Sci U S A*. 1991; 88(7):2811-4.

[34] Garry, P.S., Ezra, M., Rowland, M.J., Westbrook, J., Pattinson, K.T. The role of the nitric oxide pathway in brain injury and its treatment--from bench to bedside. *Exp Neurol.* 2015; 263:235-43.

[35] Mongin, A.A., Dohare, P., Jourd'heuil, D. Selective vulnerability of synaptic signaling and metabolism to nitrosative stress. *Antioxid Redox Signal.* 2012; 17(7):992-1012.

[36] Russwurm, M., Russwurm, C., Koesling, D., Mergia, E. NO/cGMP: the past, the present, and the future. *Methods Mol Biol.* 2013; 1020:1-16.

[37] Rauhala, P., Andoh, T., Chiueh, C.C. Neuroprotective properties of nitric oxide and S-nitrosoglutathione. *Toxicol Appl Pharmacol.* 2005; 207(2 Suppl):91-5.

[38] Sultana, R., Poon, H.F., Cai, J., Pierce, W.M., Merchant, M., Klein, J.B., et al. Identification of nitrated proteins in Alzheimer's disease brain using a redox proteomics approach. *Neurobiol Dis.* 2006; 22(1):76-87.

[39] Hill, B.G., Dranka, B.P., Bailey, S.M., Lancaster, J.R. Jr., Darley-Usmar, V.M. What part of NO don't you understand? Some answers to the cardinal questions in nitric oxide biology. *J Biol Chem.* 2010; 285(26):19699-704.

[40] Jelenković, A. Proteinopathy in Alzheimer's diseases. Belgrade, Helenn J; 2011; pp. 1-141 (In Serbian).

[41] Moncada, S., Bolaños, J.P. Nitric oxide, cell bioenergetics and neurodegeneration. *J Neurochem.* 2006; 97(6):1676-89.

[42] Bolaños, J.P., Delgado-Esteban, M., Herrero-Mendez, A., Fernandez-Fernandez, S.L., Almeida, A. Regulation of glycolysis and pentose-phosphate pathway by nitric oxide: impact on neuronal survival. *Biochim Biophys Acta.* 2008; 1777(7-8):789-93.

[43] Sarti, P., Forte, E., Giuffrè, A., Mastronicola, D., Magnifico, M.C., Arese, M. The chemical interplay between oxide and mitochondrial cytochrome c oxidase: reactions, effectors and pathophysiology. *Int J Cell Biol.* 2012; 2012:571067. doi: 10.1155/2012/571067. Epub 2012 Jul 1.

[44] Aoyama, K., Matsubara, K., Fujikawa, Y., Nagahiro, Y., Shimizu, K., Umegae, N., et al. Nitration of manganese superoxide dismutase in cerebrospinal fluids is a marker for peroxynitrite-mediated oxidative stress in neurodegenerative diseases. *Ann Neurol.* 2000; 47(4):524-7.

[45] Contestabile, A., Monti, B., Contestabile, A., Ciani, E. Brain nitric oxide and its dual role in neurodegeneration/neuroprotection: understanding molecular mechanisms to devise drug approaches. *Curr Med Chem.* 2003; 10(20):2147-74.

[46] Valko, M., Leibfritz, D., Moncol, J., Cronin, M.T., Mazur, M., Telser, J. Free radicals and antioxidants in normal physiological functions and human disease. *Int J Biochem Cell Biol.* 2007; 39(1):44-84.

[47] Hausladen, A., Stamler, J.S. Nitrosative stress. *Methods Enzymol.* 1999; 300:389-95.

[48] Ramalingam, M., Kim, S.J. Reactive oxygen/nitrogen species and their functional correlations in neurodegenerative diseases. *J Neural Transm (Vienna).* 2012; 119(8):891-910.

[49] Hannibal. L. Nitric oxide homeostasis in neurodegenerative diseases. *Curr Alzheimer Res.* 2015 Sep 20. [Epub ahead of print].

[50] Scheper, W., Hoozemans, J.J. The unfolded protein response in neurodegenerative diseases: a neuropathological perspective. *Acta Neuropathol.* 2015; 130(3):315-31.

[51] Steinert, J.R., Chernova, T., Forsythe, I.D. Nitric oxide signaling in brain function, dysfunction, and dementia. *Neuroscientist.* 2010; 16(4):435-52.
[52] Paul, V., Ekambaram, P. Involvement of nitric oxide in learning and memory processes. *Indian J Med Res.* 2011; 133:471-8.
[53] Malinski, T. Nitric oxide and nitroxidative stress in Alzheimer's disease. *J Alzheimers Dis.* 2007; 11(2):207-18.
[54] Domek-Łopacińska, K.U., Strosznajder, J.B. Cyclic GMP and nitric oxide synthase in aging and Alzheimer's disease. *Mol Neurobiol.* 2010; 41(2-3):129-37.
[55] Aquilano, K., Baldelli, S., Ciriolo, M.R. Glutathione: new roles in redox signaling for an old antioxidant. *Front Pharmacol.* 2014 Aug 26; 5:196. doi: 10.3389/fphar.2014.00196.
[56] Johnson, W.M., Wilson-Delfosse, A.L., Mieyal, J.J. Dysregulation of glutathione homeostasis in neurodegenerative diseases. *Nutrients.* 2012; 4(10):1399-440.
[57] Kolesnik, B., Palten, K., Schrammel, A., Stessel, H., Schmidt, K., Mayer, B. et al. Efficient nitrosation of glutathione by nitric oxide. *Free Radic Biol Med.* 2013; 63:51-64.
[58] Morris, G., Anderson, G., Dean, O., Berk, M., Galecki, P., Martin-Subero, M., et al. The glutathione system: a new drug target in neuroimmune disorders. *Mol Neurobiol.* 2014; 50(3):1059-84.
[59] Dasuri, K., Zhang, L., Keller, J.N. Oxidative stress, neurodegeneration, and the balance of protein degradation and protein synthesis. *Free Radic Biol Med.* 2013; 62:170-85.
[60] Konig, J.F.R., Klippel, R.A. The rat brain. A stereotaxic atlas of the forebrain and lower parts of the brain stem. Baltimore: The Williams and Wilkins Company; 1963.
[61] Gurd, JW., Jones, L.R., Mahler, H.R., Moore, W.J. Isolation and partial characterization of rat brain synaptic plasma membranes. *J Neurochem.* 1974; 22:281-90.
[62] Villacara, A., Kumami, K., Yamamoto, T., Mršulja, B.B., Spatz, M. Ischemic modification of cerebrocortical membranes: 5-hydroxytriptamine receptors, fluidity, and inducible in vitro lipid peroxidation. *J Neurochem.* 1989; 53:595–601.
[63] Anderson, M.E. Tissue glutathione. In: Greenwald, R.A. (ed.). CRC handbook of methods for oxygen radical research. Boca Raton: CRC Press; 1985. pp. 317-323.
[64] Ward, R.J., Zhang, Y., Crichton, R.R. Aluminium toxicity and iron homeostasis. *J Inorg Biochem.* 2001; 87:9-14.
[65] Stevanović, I.D., Jovanović, M.D., Colić, M., Jelenković, A., Bokonjić, D., Ninković, M., et al. N-nitro-L-arginine methyl ester influence on aluminium toxicity in the brain. *Folia Neuropathol.* 2011; 49:219-29.
[66] Stevanović, I.D., Jovanović, M.D., Jelenković. A., Ninković, M., Dukić, M., Stojanović, I., et al. The effect of inhibition of nitric oxide synthase on aluminium-induced toxicity in the rat brain. *Gen Physiol Biophys.* 2009; 28 Spec No:235-42.
[67] Stevanović, I.D., Jovanović, M.D., Jelenković, A., Colić, M., Stojanović, I., Ninković, M. Effects of L-NAME, a non-specific nitric oxide synthase inhibitor, on AlCl3-induced toxicity in the rat forebrain cortex. *J Vet Sci.* 2009; 10:15-22.
[68] Vučetić-Arsić, S., Radonjić, N.V., Jovanović, M., Selaković, V., Nikolić, T., Velimirović, M., et al. Oxidative stress precedes mitochondrial dysfunction in gerbil brain after aluminum ingestion. *Environ Toxicol Pharmacol.* 2013; 36:1242-52.
[69] Hu, Y., Zhu, D.Y. Hippocampus and nitric oxide. *Vitam Horm.* 2014; 96:127-60.

[70] Navarro-Gonzálvez, J.A., García-Benayas, C., Arenas, J. Semiautomated measurement of nitrate in biological fluids. *Clin Chem.* 1998; 44:679-81.

[71] Stevanović, I.D., Jovanović, M.D., Colić, M., Ninković, M., Jelenković, A., Mihajlović, R. Cytochrome c oxidase activity and nitric oxide synthase in the rat brain following aluminium intracerebral application. *Folia Neuropathol.* 2013; 51(2):140-6.

[72] Jelenković, A., Jovanović, M.D., Petronijević, N. The Protective effects of intrahippocampal application of green tea leaf extract on aluminium-induced brain toxicity. In: Powell, N. (ed.). *Green tea and health: antioxidant properties, consumption and role in disease prevention.* New Yourk; Nova Publishers; 2015; pp. 33-56.

[73] Bielarczyk, H., Jankowska, A., Madziar, B., Matecki, A., Michno, A., Szutowicz, A. Differential toxicity of nitric oxide, aluminum, and amyloid beta-peptide in SN56 cholinergic cells from mouse septum. *Neurochem Int.* 2003; 42(4):323-31.

[74] Mujika, JI., Ugalde, J.M., Lopez, X. Aluminum interaction with glutamate and α-ketoglutarate: a computational study. *J Phys Chem B.* 2014; 118(24):6680-6.

[75] Tórtora, V., Quijano, C., Freeman, B., Radi, R., Castro, L. Mitochondrial aconitase reaction with nitric oxide, S-nitrosoglutathione, and peroxynitrite: mechanisms and relative contributions to aconitase inactivation. *Free Radic Biol Med.* 2007; 42(7):1075-88.

[76] Zatta, P., Lain, E., Cagnolini, C. Effects of aluminum on activity of Krebs cycle enzymes and glutamate dehydrogenase in rat brain homogenate. *Eur J Biochem.* 2000; 267:3049-55.

In: Aluminum Neurotoxicity
Editor: Ankica Jelenković
ISBN: 978-1-63484-739-1

Chapter 5

ASSAIL OF ALUMINUM IN REFERENCE TO NEUROLOGICAL DISORDERS

Shaik Afsar Pasha, Ch. Sindhu and Prasunpriya Nayak*
Departments of Physiology and [1]Neurology,
NRI Medical College and General Hospital,
Chinakakani, Mangalagiri Mandal, Guntur District,
Andhra Pradesh, India

ABSTRACT

Aluminum is a well known neurotoxin. There are many reports in support and against of its involvements in different neuropathological conditions. Focusing towards the reports of acute and chronic exposures of aluminum, here efforts are made to present the symptoms of aluminum intoxication and their logical correlation with affected structures. In addition, neurological correlates between aluminum exposure and pathophysiology of Alzheimer's diseases (AD), Parkinson's disease (PD), Huntington's disease (HD), Autism spectrum disorders (ASD), Amyotrophic lateral sclerosis (ALS), Gulf war syndrome (GWS) and Fahr's diseases are discussed. Not only limiting to evaluation of neurological disorders with cognitive impairment, aluminum related problems should be explored in neuropathologies having components of motor disorders.

Keywords: aluminum, neurotoxicity, Alzheimer's diseases (AD), Parkinson's disease (PD), Huntington's disease (HD), autism spectrum disorders (ASD), amyotrophic lateral sclerosis (ALS), Gulf war syndrome (GWS), Fahr's diseases

---

* Dr. Prasunpriya Nayak, Professor of Physiology, NRI Medical College & General Hospital, Chinakakani, Mangalagiri Mandal, Guntur District, Andhra Pradesh 522503. India. Email: nprasunpriya@gmail.com.

Introduction

An invigorated necrobiosis due to chronic perturbed metabolic activities is usually seen in case of aluminum toxicity (Metwally and Mazhar, 2007). Though the brain is not a favored target organ as per the toxicodynamics of aluminum in humans (Skalasky and Charchman, 1983), heightened exposure along with compromised in-built protective measures for the brain and allostasis in aluminum toxicodynamics may allow aluminum to gain access to this aluminum-vulnerable organ (Nayak et al. 2011; Nayak, 2012). Inveterate exposures, even in minuscule quantities, favor bioaccumulation of aluminum in brain tissues and portray neuropathological traits of Alzheimer's disease (AD) (Shaw and Marler, 2013). Even after long standing debate regarding the role of aluminum in AD, current advisement cannot rule out the aluminum hypothesis for AD (Lidsky, 2014). Neurotoxicity of some extent is an obvious fact consociated with aluminum exposure, and interaction of aluminum with many biomolecules opens up the risk of different associated pathological consequences. Enhanced degenerative changes in both neuronal and glial cells were found to be associated with presence of aluminum in their microenvironment (Gupta et al. 2005). It is likely that several physical, chemical and physiological factors decide the persuasiveness of aluminum's impact on these cells. Accordingly, multifarious effects of aluminum have been reported in the brain (Yokel, 2013). Even though, there are controversies about the role of aluminum in neuronal diseases, their concurrence supports mostly assertive. Presence of aluminum has been reported in several neurological disorders (Mujika et al. 2011; Nayak, 2012) with varied neurological symptoms (Crapper and De Boni, 1980; Zatta et al. 1991, Malaki, 2013); list of these disorders (Table 1) is gradually increasing (Zatta et al. 1991; Kawahara et al. 2001; Sánchez-Iglesias et al. 2009; Bondy, 2010).

Table 1. Maladies associated with aluminum

Neurological Disorders	*Acute Symptoms*
Alcohol dementia with patchy demyelination, Alzheimer-type senile and presenile dementia, Amyotrophic lateral sclerosis (spinal cord and brain), Autism spectrum disorders, Chronic fatigue syndrome, Down syndrome with manifested Alzheimer's disease, Encephalopathy and β_2-microglobulin-associated amyloidosis in dialysis patients, Myalgic encephomyelitis Neurofibrillary degeneration adjacent to harmatoma Neural tube defect (new born) Parkinsonian dementia with neurofibrillary degeneration, Parkinsonism dementia (Kii Peninsula of Japan and Guam), Senile plaques of Alzheimer's disease, Straitonigral syndrome.	Confusion Delirium Dis-orientation Hallucinations Slurred speech Tremors.
Non-neurological disorders	*Chronic Symptoms*
Aluminosis. Aluminum fibrosis. Crohn's disease. Macrophagic myofasciitis. Microcytic anemia without iron deficiency. Osteomalacia.	Ataxia, Dysarthria, Dysphasia, Dyspraxia, General weakness, Generalized convulsions with status epilepticus Impaired coordination, Jerking movements (myoclonic jerks), Loss of curiosity, Memory loss, Sluggish motor movement, Tremor.

Toxicity, including neurotoxicity, of aluminum is known to human for more than a century (Tomljenovic, 2011). Because of utility-friendly physical characteristics and ostensible 'biologically inert' physico-chemical properties of aluminum, its use has become so widespread that its toxicity is overlooked and overleaped by its uses during the so called 'Aluminum age'. Serious concerns about the aluminum-mediated toxicity started growing nearly fifty years ago and exhaustive researches have been being carried out for last two to three decades. Multidimensional studies with molecular level approach revealed a number of probable mechanisms of aluminum-induced cellular desolation (Nayak et al. 2015). Precise contributions of these aluminum-bourne impacts (Table 2) in aluminum-induced neurodegenerative changes reported in neuronal and glial cells are unclear and obnubilating reports suggest involvement of more than one mechanism in the process; nevertheless, tissue specific divergences in the course of ravage cannot be ruled out.

Table 2. Proposed targets of aluminum's assail

Amyloid pathway	Membrane transport
Apoptotic regulators	Metal binding proteins
Intracellular [Ca^{2+}]	Nucleotides
Cell-cell interaction	Neurofilaments
Lipid metabolism	Neuroinflammation
Energy metabolism	Neurotransmitters
Genetic machinery	NO - system
Immunomodulators	Oxidative stressors

Interestingly, aluminum preferentially exploited apoptotic pathways and necrotic pathways to affect the viability of neurons and glial cells, respectively, as indicated by in vitro studies (Zhang et al. 2008). Non-renewability of neuronal cells and high rate of renewability in glial cells would have been helped astrocytes to overcome the likely eventuality of preceding astrocyte death in comparison to neuronal death during neurodegenerative upshots (Bharathi et al. 2008). Considering the importance of glial cells as a support system for the neuron, furthering of apoptotic changes in neurons can be anticipated in case the rate of astrocytic necrobiosis outnumbers its regeneration (Zhang et al. 2008). Nevertheless, aluminum-induced excitotoxicity as well as apoptosis in astrocytes are also reported (Aremu et al. 2011; Blaylock, 2012).

Disregarding the mode and diverseness of exposure, aluminum is invariably neurotoxic for all types of experimental animals (Liang et al. 2012). However, overt developmental consequences are only observed in animals with high doses of perinatal exposure to aluminum (Yumoto et al. 2001; Nayak, 2006); humans are also no exception.

Symptoms of Aluminum's Assail on Nervous System

It is rightly mentioned by Exley (2014) that chronic aluminum intoxication wouldn't have any specific identifiable symptom(s) and implied that the indication of aluminum toxicity can be multifaceted as well as in disguise of other symptoms. Neurological problems because of aluminum are known for long time; however, confirmation of its neurodegenerative impacts was by identification of a link between use of aluminum as phosphate binder in dialysis fluid and development of dialysis encephalopathy. Incidentally, in 1962, McLaughlin et al. identified two important facts about the aluminum intoxication – (a) for most, aluminum was found to be harmless and acting as antidote for free silica, while some others with certain conditions were developing pulmonary fibrosis upon long exposure to aluminum; (b) few of those developing aluminum-induced pulmonary fibrosis exhibit neurological problems (McLaughlin et al. 1962). Recently, Gherardi et al. (2015) echoed the same problem for use of alum as adjuvants and suggested the genetic susceptibility could be important factor for the reported intolerance to aluminum in certain individuals.

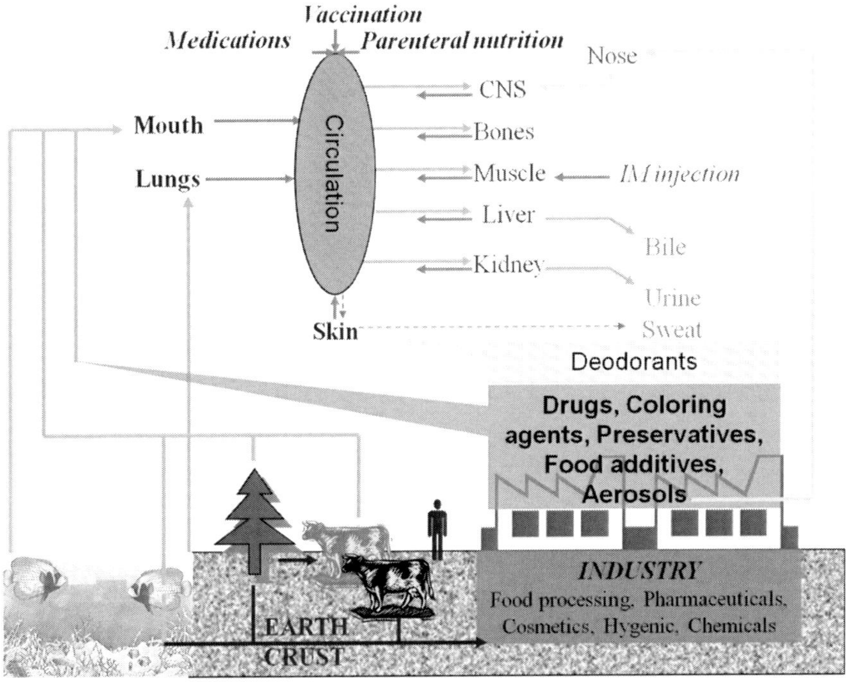

Figure 1. Exposures to aluminum and its tissue distribution (Nayak, 2014).

Diversity of exposure routes allows aluminum to accumulate in different organs (Figure 1). Brain and lungs are the major target organs for inhalation exposure to aluminum. Exposure via vaccination or intramuscular injection targets muscles as aluminum is found in high concentrations as an adjuvant in various vaccines (Shaw et al. 2014; Gherardi et al. 2015). Liver and stomach are specifically targeted for oral aluminum exposure. Similarly, external application of aluminum is likely to target skin as an accumulation site. However, bones, kidneys, reproductive organs accumulate aluminum even without being involved as route of aluminum exposure, except in case of rare iatrogenic exposure. Interestingly, most concerned and sought-after clinical features are neurological signs and symptoms. Clinically, there are no specific signs and symptoms ascribed to be diagnostic or confirmative of aluminum intoxication, however, excessive headaches, depression, numbness of proximal limbs, blurred vision and abnormal heart rhythm are commonest features associated with aluminum toxicity. Functionally, impairment in choice reaction time, long-term memory, psychomotor speed and recall are also found to be associated with aluminum intoxication (Life Sources, http://www.life-sources.com/pdf/toxic-heavy-metals-booklet.pdf). This indicates that both central and peripheral nervous tissues are potential target organs irrespective of the route of exposure.

Commonly, the symptoms for aluminum intoxication are presented and looked for only in case of suspected chronic aluminum exposure. However, symptoms of acute exposure may also be helpful to correlate with. In a case of acute aluminum exposure in 18 month old boy with congenital nephrotic syndrome developed head tics and grimace on the 4th day of use of aluminum hydroxide (150mg/kg/day) as phosphate binders (Malaki, 2013). On the 5th day, the symptoms progressed to declining cognition (unresponsive to mothers face and voice), head tremor, incoherent speech progressing to speechless state. Authors reported that the symptoms resolved after discontinuation of drug and continuous dialysis over 1 month (Malaki, 2013). The case report suggests possibilities of neurological symptoms of aluminum exposure even with only 5 days of acute iatrogenic exposure directly to blood.

Sińczuk-Walczak et al. (2003) reported increased emotional irritability, headache, concentration difficulty, sleep disorders, vertigo, mood lability, anxiety and fear as subjective CNS symptoms pronounced by the majority of studied foundry workers exposed to Al_2O_3 in concentrations ranging from 0.13 to 19.5 mg/m^3 during aluminum production and found that some of these complaints were significantly associated with duration of exposure, however, not to current urinary or ambient concentration.

In a recent cross-sectional analytical study, Moussa et al. (2013) demonstrated that workers from an aluminum manufacturing company having higher serum and urinary aluminum, compared to those of their unexposed relatives, had lower scores in Mini Mental State Examination while higher scores in Wechsler Memory Scale (information, orientation, logical memory, digit span and associate learning), Eysenck Personality Questionnaire (psychotism, neurotism, extroversion, lying and criminality), Hamilton Anxiety Rating Scale and Hamilton Depression Rating Scale along with significant correlation between these psychological deficits and serum-urinary aluminum level. This report is clearly evincing detrimental impact of aluminum exposure on the mental health of workers exposed to aluminum for more than 10 years, despite the lack of any report of clinical neurological complaints. Demonstrating detrimental impact of aluminum exposure on specific types of cognitive functions, Akila et al. (1999) emphasized the importance of choosing a correct test task to evaluate the cognitive functions. Reporting cognitive impairments in dialysis encephalopathy patients, English (1978) indicated that prior to overt clinical features, impaired intellectual function was noticeable for many patients, though not all.

Hammond (2006) also reviewed the cognitive impairments in miners exposed to aluminum dust and also cited article showing more slow-wave (delta and theta) electroencephalographic (EEG) activity and less alpha activity frontally in welders with high serum aluminum in comparison to the welders having low serum aluminum and provided a neurophysiological link between aluminum exposure and functional deficits in brain. Earlier, changes in EEG pattern were also noted in a group of aluminum welders and extents of alterations were grossly found to be associated with level of aluminum exposure (Riihimäki et al. 2000). However, Iregren et al. (2001) with visual and quantitative analyses did not find significant differences in EEG recorded from three different types of aluminum exposures - smelters, flake powder workers, welders and that of control. Contrary to this, more frequent alterations in EEG in a group of aluminum exposed individuals with complaints of sleepiness and vertigo were found by Sińczuk-Walczak et al. (2003). Impacts of aluminum on the neuronal functions may also depend on susceptibility of an exposed individual, as hypothesized by McLaughlin et al. (1962). On the other hand, even if modest exposure to aluminum is not able to persuade a clinically detectable impairment in nervous tissues, possibilities of asymptomatic damage of nervous system is always there, as suggested by Sińczuk-Walczak et al. (2003).

Apart from these cognitive and affective functions, exposures to aluminum also aggress motor functions. When the urinary aluminum concentration approximates 50 μg/L level, motor functions may face insidious encumbrances in welders or other workers exposed to aluminum (Sjogren et al. 1996).

Subtle behavioral changes are noted in certain species soon after aluminum infusion. As the duration of exposure continues, the behavioral changes become more overt and severe neuropathological signs start appearing. In experimental studies, most authors accord a similar progression of symptoms (Figure 2) in aluminum-exposed subjects (Crapper et al. 1973; Selkoe et al. 1979; Crapper and Dalton, 1973).

Of late, cellulitis, seizures, fatigue, depression, pain, and death as such have been linked to exposure to aluminum through vaccination (Seneff et al. 2012). Chronic fatigue syndrome (CFS) comprising weakness malaise, speech disorders, poor concentration attention deficits having response inhibition, poor memory (short term verbal and auditory), pre-senile and senile dementia, stupor, decreased locomotor activity and convulsions (seizure) have been linked to aluminum exposure. It is suggested that removal of aluminum may improve the condition (Life Sources, http://www.life-sources.com/pdf/toxic-heavy-metals-booklet.pdf).

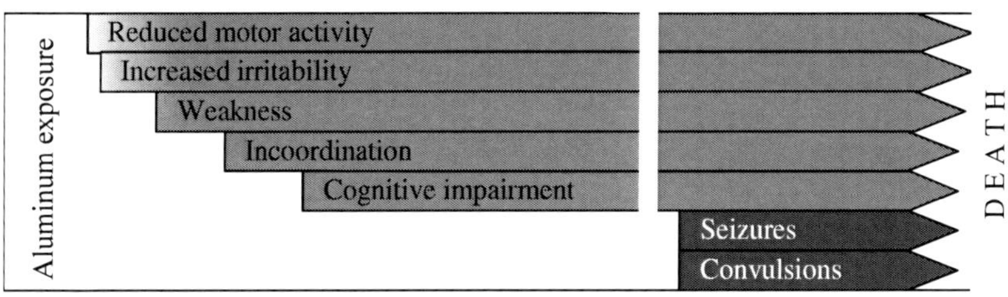

Figure 2. Sequence of symptoms of aluminum exposure in susceptible subjects.

Structural Changes Following Aluminum Accumulation

Enduring deterioration of neurological operates is reported along with generalized aluminosis and the autopsy study revealed patchy demyelination in the brain (Lapersle et al. 1975). Vitiated neuronal density along with degenerative changes like pyknosis, vacuolation, and chromatin condensation are the trenchant nuclear morphological variations in rats exposed to aluminum through drinking water (Varner et al. 1998). Aluminum-induced chromatin compaction is also reported by Lukiw (2010) and specific biological role of such compaction is suggested. Apart from these nuclear changes, Varner et al. (1998) have demonstrated that chronic administration of aluminum to rats in drinking water can cause other cytoskeletal and cerebrovascular abnormalities. Upon insult with aluminum exposure, microscopic aluminum and phosphorus compounds are also detected in the brain along with demyelination (Zatta et al. 1991).

Within the brain, aluminum-induced microscopic degenerative changes are found to be widely distributed. Along with the biochemical alterations in the frontal cortex (Nayak et al. 2013; Tripathi et al. 2009), temporal cortex (Nayak et al. 2013), parietal cortex and hippocampus, these brain regions also undergo massive cellular depletion (Nayak, 2014; Nayak et al. 2014; Sarkaki et al. 2009). This observation suggests that aluminum intoxication impairs not only the process of learning and memory, but other higher (intellectual) functions as well. Some degree of functional correlates is also observed between aluminum-induced cytoarchitectural degenerative changes and aluminum-induced impairment in neurobehavioral parameters (Nayak et al. 2014). Therefore, the impact of aluminum should be assessed in the behavioral alterations where skilful motor memory is involved through learning (training).

Biochemical alterations in the cerebellum in response to aluminum exposure were not so prominent as in other parts of the brain. Nevertheless, the degenerative changes have been intensified in the presence of pro-oxidant dominance (Nayak et al. 2011). Cerebellar atrophy along with time-related deterioration of the Purkinje cells is reported in aluminum intoxication through intracerebroventricular administration (Ghetti et al. 1985). On the other hand, Yokel (1994) indicated that aluminum-induced alterations in the cerebellar Purkinje cells are similar to those observed in aged human brains. Studies with special histological staining technique for aluminum indicated that in conditions like pro-oxidant dominance, Purkinje cells are accumulating aluminum in their processes and it is suggested to be linked with the behavioral alterations observed in rats (Nayak, 2014).

Even reckoning reported controversies regarding aluminum-induced formation of neurofibrillary tangles (NFT); the pathway of neurodegenerative changes caused by aluminum still cannot undermine the role of NFT formation. Accumulation of neurofilaments is reported in the perikaryal regions of neurons when exposed to aluminum through intra-cerebral administration (Savory et al. 1991). Accumulation of normal-appearing neurofilaments, in response to aluminum insult, is reported in dendrites and perikaryal regions of brain stem neurons (Terry and Pena, 1969). Aluminum intoxication also causes axonal swelling (Troncoso et al. 1982).

Compared to the brain, less morphometric alterations are found to be associated with spinal cord as an impact of aluminum exposure. Vacuolation in spinal cord is reported in response to chronic aluminum exposure to mice (Golub and Tarara, 1999). Upon exposure to aluminum, accumulation of NFT in spinal cord neurons is also reported by Terry and Pena (1969). Granulo-vacuolar degeneration, deposition of lipofuscin, mitochondrial swelling, thinning of myelin sheath, and increased frequency of perforated synapses in the brains of rats exposed to aluminum are reported through ultrastructural studies (Miu and Benga, 2006). Decreased thickness of postsynaptic density, increased width of synaptic cleft, increased numbers of flat synapses, decreased numbers of positive curvature synapse and perforated synapse are observed in a study of aluminum-induced learning and impairment model of rat (Shi-Lei et al. 2005).

Histochemical and immunocytochemical studies of Platt et al. (2001) suggests that the enhancement of inflammation and the interference with cholinergic projections may be the modes of action through which aluminum cause learning and memory deficits.

NEUROPATHOLOGIES ASSOCIATED WITH ALUMINUM ACCUMULATION

Mechanistic approaches to elucidate the paths of aluminum-mediated neurotoxicity opens up large and endless possibilities through which aluminum may intervene in the metabolic pathways in the brain (Chaitanya and Nayak, 2015; Exley, 2014). Aluminum is a well accepted neurotoxin and entailed in a number of neurodegenerative diseases like AD and encephalopathy (Han et al. 2013). In a study to assess brain aluminum levels in patients who died of different neurological disorders, Traub et al. (1981) have observed high levels of aluminum in seven out of seven samples of AD, two out of three samples of Huntington's disease, two out of two samples of Parkinson's disease, one out of one sample of progressive supranuclear palsy, one out of one sample of acoustic neuroma, one out of two samples of cerebrovascular disease, and one out of two samples of Guamanian amyotrophic lateral sclerosis (ALS), while nine samples of nine Creutzfeldt-Jakob disease demonstrated normal brain aluminum. Even though, mere presence of aluminum need not be evidence of its involvement in the disease process, its role in those conditions, where its presence was noted significantly higher, cannot be ruled out. Similarly Exley and House (2011) also have reviewed the localization of aluminum in neurons, NFTs, lipfuscin and senile plaques in AD; neurons and NFTs of dementia pugilistica; Lewy bodies in Parkinson's disease; lipofuscin in aged brains; pathological lesions in Fahr's and Behcet's disease; myelin in progressive leukoencephalopathy; and neurones, glia and the choroid epithelium in aluminum-induced encephalopathies.

Extensive researches including epidemiological, neuropathological, and biochemical studies have made the status of aluminum debatable for its much acclaimed causative role in the etiopathogenesis of AD. Lidsky (2014) has suggested 'aluminum hypothesis' for AD is surviving because there is no other acceptable answer, whereas, Yegambaram et al. (2015) have included aluminum in the possible environmental contaminants to instigate AD. In this context, current development in relation to the aluminum's role in other neurodegenerative disorders indicates that aluminum may not be the sole cause of specific neuropathological conditions, but its involvement in the process of neurodegeneration and neurobehavioral impairment cannot be disregarded.

As discussed previously, the experimental evidences indicate that there is some regional and/or site preferences in aluminum accumulation in the central nervous system, however, this not necessarily true for its impact in terms of neurodegenerative changes. Thippeswamy et al. (2013) have accentuated aluminum as ordinarily harmless and as inexplicitly nontoxic with a note on its presence in high concentration in people with Parkinson's disease, senile dementia, ALS, and senile dementia associated with AD. On the other hand, some age groups are specifically vulnerable to aluminum accruement in the brain (Nayak 2012, Gherardi et al. 2015). Therefore, the diseases associated with aluminum intoxication are aggrouped (Table 3) on the basis of development of the disease and regional involvement in the disease process(es).

Table 3. Grouping of aluminum-lined neuropathological conditions

Cranio-caudal localization of disease affliction Period of development of symptoms	Brain (Parenchyma)	Basal Ganglia	Spinal cord	Spinal nerves	Muscle	Non-specific sites
Degenerative diseases						
Alzheimer's disease	✓					
Parkinson`s disease	✓	✓				
Amyotrophic lateral sclerosis	✓		✓			
Fahrs disease	✓	✓				
Huntington's disease	✓	✓				
Pick's disease	✓					
Developmental diseases						
Developmental diseases of childhood (Autism spectrum disorders)	✓					
Seizures	✓					
Acquired (Inflammatory) diseases						
Multiple sclerosis	✓		✓			
Guillain barre syndrome				✓		
Behcet disease	✓	✓				✓
Macrophagic myofasciitis	✓				✓	
Acquired (Post-traumatic) diseases						
Dementia pugilistica	✓					
Other diseases						
Chronic fatigue syndrome					✓	✓
Congophilic amyloid angiopathy	✓					
Dialysis associated encephalopathy	✓					
Progressive leuco-encephalopathy	✓					
Potroom palsy	✓		✓		✓	
HIV associated dementia	✓					
Gulf war syndrome	✓		✓			
Seasonal affective disorder						✓
Viral encephalitis	✓					

Another important aspect of aluminum's role in neuropathological changes is oxidative stress (Figure 3). Being a redox inactive metal, aluminum is not directly involved in cellular oxidant homeostasis. Nevertheless, there are many important pathways through which presence of aluminum can be proven to be an imperilment for cellular oxidant balance (Han et al. 2013; Nayak, 2012; 2014). Exposures to aluminum augment the oxidative stress and reduce the superoxide-peroxide handling capacity (SPHC) of the brain in a pro-oxidative dominant condition (Chaitanya et al. 2012, Nayak et al. 2013). Along with the neuronal loss, oxidative imbalance during the progression of neurodegenerative diseases particularly ALS, AD and Parkinson's disease (Abubakar et al. 2007), provides opportunity to intensify assails of aluminum in neurons. As many of the neuropathological conditions, which are colligated to aluminum, are also associated with aging, evidence from clinical and animal model studies demonstrate that brain aluminum content increases with age which may either be due to increasing exposure with age or decreased ability to remove aluminum from the body (Markesbery et al. 1984).

Even though there are numerous neuropathological conditions which are likely to be associated with aluminum-induced neurotoxicity (Table 1), most studies related to aluminum-linked neuropathologies focused mainly on a few. Here, some of the extensively studied neurological disorders are discoursed with suggested links to aluminum toxicity.

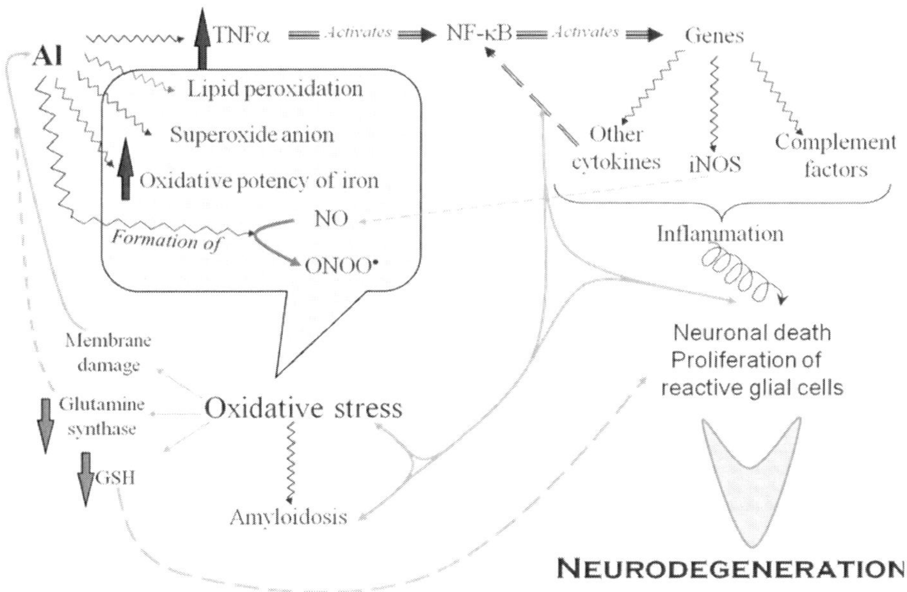

Figure 3. Suggested pathways for oxidative stress and neurodegeneration caused by aluminum exposure (Nayak, 2014).

Alzheimer's Disease

As indicated already, a direct role of aluminum has not yet been established definitively and inclusion of aluminum in the list of causative agents is still under debate, however, it's corroborative value is validated by numerous proofs. As a result of protein build up forming senile plaques and intracellular neurofibrillary tangles in the grey matter and hippocampus (thought processing and memory centers), there is progressive mental deterioration causing dementia, which includes general loss of cognitive function, memory loss and difficulties with thinking, problem solving, and language (Shaw 2013). In turn, this leads to selective loss of synaptic connections, death of nerve cells, and overall loss of brain tissue. Aluminum is reportedly playing an important role in formation of NFTs by cross-linking and inducing conformational changes in the non-toxic regular neurofibrils (Kawahara, 2011).

Oxidative stress plays a predominant role in the pathogenesis of neuronal degeneration. As aluminum exists as a trivalent cation, it is capable of enhancing iron based oxidation, which in turn promotes generation of reactive oxygen species with the help of iron, leading to oxidative stress (Campbell, 2002; Bondy, 1998; Nayak, 2014). Accumulating amyloid-β also induces oxidative stress by interpolating into neuronal membranes, and producing lipid peroxidation, protein oxidation, and formation of reactive oxygen and nitrogen species (Butterfield et al. 2010). Therefore, in the presence of aluminum, possibly a vicious cycle of neurodegeneration is initiated including aluminum, beta amyloid plaque and oxidative stress. It is most likely that it is going to develop a pro-oxidant dominant condition in the neuronal microenvironment, the neuropathological peril of aluminum will be further supported.

AD is associated with elevated basal levels of markers for inflammation; aluminum being a subtle promoter. Aluminum salts can increase levels of glial activation, inflammatory cytokines, and amyloid precursor protein in the brain (Bondy, 2010). It also inhibits cholinergic functioning and synaptic uptake of dopamine, norepinephrine, 5-hydroxytryptamine, suggesting aluminum is implicated in other various neurodegenerative disorders.

Parkinson's Disease

Although there is currently little data on the potential impact of aluminum and Parkinson's disease (PD), aluminum in cell cultures have been shown to be conducive to some of the same mis-folded protein accumulations as seen in the human PD (Uversky et al. 2001). There are many associated links which suggest a plausible connection between PD and aluminum exposure. Aluminum in the whole substantia nigra appeared to increase in patients, significantly or not, in comparison with the controls (Rajput et al. 1985; Uitti et al. 1989). In the globus pallidus as a whole and in the *gyrus hippocampalis*, the unique datum retrieved for aluminum indicated a significant enrichment in PD patients. However, the report of significant aluminum elevation found in *nucleus caudatus* of PD by Yasui et al. (1992) was not confirmed (Speziali and Orvini, 2003). Apart from these, Gurruto et al. (1984) also reported accumulation of aluminum in Parkinsonism-dementia of Guam.

Even in the presence of confusing reports, the possibilities of aluminum hindrance in the formation of neuropathological changes like that of PD cannot be overruled. In this regard, oxidative stress (including dopa-metabolizing pathways) and α-synuclein has received lot of attention from the researchers and indicated affirmative possibilities (Figure 4).

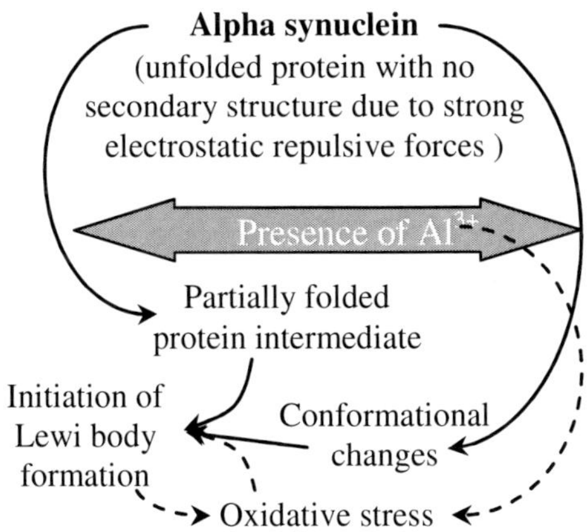

Figure 4. Possible role of aluminum in the pathogenesis of Parkinson's disease.

Huntington's Disease

Even though there is no direct evidence of aluminum's involvement in the Huntington's disease (HD), likeliness of aluminum-induced aggravation or even precipitation of HD through development of oxidative stress (Yu et al. 2014) or excitotoxicity (Winter and Winter, 2007) cannot be ruled out. In support of this view, Traub et al. (1981) already reported the high level of aluminum in brains of two patients died with HD.

Autism Spectrum Disorder

Aluminum is also implicated in autism spectrum disorders (ASD), albeit no direct link has been established yet between them. However, there are occasions when ASD and aluminum exposures coexist. Autistic patients are vulnerable to toxic metals due to insufficient serum sulfate, glutathione, cysteine, cholesterol sulfate, and sulfate (Seneff et al. 2012). Oxidative stress related biomarkers have identified consistent deficiency in glutathione in autism (Seneff et al. 2012); which is a potential link for aluminum to be involved in the progression of the disease process by augmenting the oxidative stress, as suggested in animal experimentation (Nayak et al. 2010). A study on 28 individuals with ASD diagnosis have shown their levels of free sulfate in the bloodstream was only 33% of that of the controls subjects; Seneff et al. (2012) hypothesized that this would result in insufficient ionic buffering in the vasculature, with inadequate sulfation of extracellular matrix proteins that are essential for proper colloidal suspension of particles and cells. Significance of serum taurine, a sulfur containing amino acid derived from cysteine, in amelioration of aluminum induced hepatic injury has been suggested (Yeh et al. 2009). Enhanced activities of endogenous antioxidants evidenced by the significant increase in hepatic reduced glutathione level and activities of glutathione peroxidase and catalase, is the proposed mechanism for such activity (Yeh et al. 2009). On the other hand, the inherent mechanism of aluminum disposal (Nayak, 2012) is also likely to be impedemented as taurine is depleted in ASD due to insufficient cysteine and reduction in bile acids (Seneff et al. 2012; Yeh et al. 2009).

Aluminum is found in high concentrations as an adjuvant in various vaccines (Shaw et al. 2014; Gherardi et al. 2015). Autistic children also have poorly developed barrier function in epidermis so aluminum can easily penetrate the skin (Seneff et al. 2012). Therefore, vaccination of autistic children with aluminum adjuvant raises the body aluminum burden by many folds and may perpetuate the neuropathological conditions as an opportunistic peril. Blood brain barrier permeability, inflammation, deranged calmodulin signaling, impaired astrocyte function, interference with iron-sulfur clusters in mitochondria resulting in impaired ATP synthesis, excess production of ROS, and irregular formation of cytoskeleton - are some of the possible occasions where aluminum may hasten the pathological aggress in ASD (Nayak, 2012; Seneff et al. 2012; Shaw, 2013). Autistic and immune-compromised children are thereby more susceptible to the effects of aluminum toxicity – damage and rupture of phagolysosomes, generation of reactive oxygen species (Seneff et al. 2012).

Amyotrophic Lateral Sclerosis

Amyotrophic lateral sclerosis (ALS) is a progressive disease of still unknown origin that targets the motor neurons in the brain, brainstem and ventral spinal cord. Typically, at end-stage disease, both sets of motor neurons have undergone degeneration with resulting loss of motor function. Death typically occurs by respiratory failure. The typical age for the onset of ALS starts is mid-50 s to 70 s, and the survival time after diagnosis ranges from 3 to 5 years.

Many ALS victims show a significant loss of cognitive function as well at the latter stages of the disease.

About 90% of all ALS cases (sporadic ALS) arise from unknown factors, while 10% are 'familial' with a variety of genes involved, notably mutations in the genes coding for the protein superoxide dismutase. Of the 90% of sporadic cases, a current view is that environmental toxins, alone or in synergy with still unknown susceptible genes, are to blame. Even though the details of suggested environmental toxin remain unidentified (Shaw et al. 2008); aluminum is a potential candidate for the same.

Aluminum is likely to cause selective damage to the motor neuronal pool of spinal cord either directly or indirectly (Shaw and Petrik, 2009). Accumulation of aluminum into the cytoplasm via cellular uptake mechanisms or diffusion could cause alterations in glutaminase and glutamine synthetase and thus alter the availability of the neurotransmitter glutamate. In addition, aluminum-induced abnormal 'tau' protein accumulation could increase NFTs and impair cellular transport mechanisms. Aluminum may alter the synaptic structures also – by decreasing the thickness of post-synaptic density, by increasing the width of the synaptic cleft or by increasing the number of flat synapses. Additionally, aluminum exposure can also block synaptic functions through voltage-activated calcium channels, augmented acetylcholinesterase activity or by its accumulation at the synaptic cleft (Shaw and Petrik, 2009). Therefore, ample opportunities for aluminum to alter the inter-neuronal communications, which may be implicated in the etiopathology of ASD. Apart from these, as discussed earlier, aluminum has possibilities to alter the neuron-glial relationship by inducing apoptosis in astrocytes (Aremu et al. 2011; Aremu et al. 2006).

Indirect toxicity of aluminum could occur by activating various cytokines (Johnson et al. 2003), releasing glutamate in an excitotoxic cascade, or by modifying various enzymatic pathways (Nayak and Chatterjee, 2001) and by stimulating abnormal, generalized immune responses.

Gulf War Syndrome

Among the three key features of Gulf War Syndrome (GWS), neurological symptoms include blurred vision, concentration and memory problems, irregular heartbeat, loss of balance and dizziness, speech difficulties, sudden loss of strength, tremors, shaking, and musculoskeletal/rheumatologic symptoms including generalized muscle aches, joint aches, numbness in the hands and feet, and swelling in the joints and in the extremities, are likely to be associated with aluminum intoxication (ADF, 2015). Many of these symptoms could be linked to reports of aluminum exposure to human (occupational) or animal (experimental).

Fahr's Disease

Fahr's disease has a broad spectrum of clinical neurological presentation. Patients can be presented with movement disorders, executive and cognitive impairment, seizures, pain, pyramidal symptoms, cerebellar signs depending on part of the brain affected by lesion (Manyam et al. 2005). The most common sites of calcifications are basal ganglia and dentate nuclei of cerebellum. However, calcification can involve centrum ovale, thalami, cerebral cortex, sub-cortex and no part of the brain could be spared. The extension of calcification is directly related to the psychiatric manifestations, but not to the neurological complaints.

Deposits are composed of mineral compounds like calcium phosphate and carbonate; glyconate, mucopolysacchride and metals including aluminum (Bouras et al. 1996).

CONCLUSION

Neurotoxicity of aluminum is widespread. Because of unavailability of its appropriate alternative and omnipresence in every spheres of life, exposures to aluminum are unavoidable. There is no biological need of aluminum in any life form; hence, the presence of aluminum is unwanted. As aluminum can interact with many biological systems, physiological performances of these baffled systems are vulnerable to be influenced by aluminum intoxication. Oxidative stress is found to augment the neurotoxicity of aluminum, while aluminum is likely to initiate oxidative stress in certain conditions; thus, a vicious cycle is perpetuated involving aluminum, oxidative stress and their neurodegenerative impacts. Even though, the involvement of aluminum is still debatable in neurological disorders, in all likelihood, aluminum is having some covert impacts either directly or through oxidative stress. Mostly, the neurological diseases having components of cognitive impairment are appraised for aluminum's involvement, however, the tendency seen of late and experimental evidences indicate that scope should be widened to include the neurological diseases having elements of motor disorders. Whether aluminum's role in the neurological disorder is established or not, efforts should be made to minimize the exposure to aluminum.

REFERENCES

Abubaker, M, A. Taylor, A Ferns. 2007. Effects of Aluminium Toxicity on Primary Cortical Astrocytes. *The Internet Journal of Toxicology* 4(2). Accessed August 8, 2015.

ADF (The Alzheimer's Diseases Fund) Aluminum in Alzheimer's Disease and Other Neurological Disorders. www.alz-disease.org/downloads/Aluminum1.pdf Accessed November 10, 2015.

Akila, Ritva, Stollery Brian T, and Riihimaki Vesa. 1999. Decrements in Cognitive Performance in Metal Inert Gas Welders Exposed to Aluminium. *Occupational and Environmental Medicine* 56:632-39.

Aremu, David A, Ezomo Ojeiru F, and Meshitsuka Shunsuke. 2011. Gene Expression in Primary Cultured Astrocytes Affected by Aluminum: Alteration of Chaperons Involved in Protein Folding. *Environmental Health and Preventive Medicine* 16:16-24.

Aremu, David, and Meshitsuka Shunsuke. 2006. Some Aspects of Astroglial Functions and Aluminum Implications for Neurodegeneration. *Brain Research Reviews* 52:193-200.

Bharathi, VP, M. Govindaraju, AP Palanisamy, K. Sambamurthi, and KSJ Rao. 2008. Molecular Toxicity of Aluminum in Relation to Neurodegeneration. *Indian Journal of Medical Research* 128(4):545-56.

Blaylock, Russell L. 2012. Aluminum Induced Immunoexcitotoxicity in Neurodevelopmental and Neurodegenerative Disorders. *Current Inorganic Chemistry* 2(1):46-53.

Bondy, SC, SX Guo-Ross, and J. Pien. Mechanisms Underlying Aluminum-induced Potentiation of Oxidant Properties of Transition Metals. *Neurotoxicology*, 1998, 65-72.

Bondy, Stephen C. 2010. The Neurotoxicity of Environmental Aluminum is Still an Issue. *Neurotoxicology* 31:575-81.

Bouras C, P. Giannakopoulos, P. F. Good, A. Hsu, P. R. Hof, and D. P. Perl. 1996. A Laser Microprobe Mass Analysis of Trace Elements in Brain Mineralizations and Capillaries in Fahr's Disease. *Acta Neuropathologica* 92(4):351-7.

Butterfield D Allan, V. Galvan, M. B. Lange, H. Tang, R. A. Sowell, P. Spilman, J. Fombonne, O. Gorostiza, J. Zhang, R. Sultana, and Bredesen DE. 2010. In vivo Oxidative Stress in Brain of Alzheimer Disease Transgenic Mice: Requirement for Methionine 35 in Amyloid Beta-Peptide of APP. *Free Radical in Biology and Medicine* 48:136–144.

Campbell Arezoo. 2002. The Potential Role of Aluminium in Alzheimer's Disease. *Nephology Dialysis Transplantation* 17(Suppl 2):17-20.

Chaitanya Balasai, TV, and P. Nayak. 2015. Influences of Exogenous Pro- and Anti-oxidants on Aluminum-induced Behavioral Alterations in Elevated plus Maze and Passive Avoidance Activity of Rats. *Scholars Academic Journal of Biosciences* 3(5):434-42.

Chaitanya Balasai, TV, M Kalyani, JS Bondili and P Nayak. 2012. Effect of Aluminum Exposure on Superoxide and Peroxide Handling by Liver, Kidney, Testis and Temporal Cortex of Rat. *Indian Journal of Biochemistry and Biophysics* 49(5):395-8.

Crapper D. R., and U. De Boni. 1980. Aluminum. In: *Experimental and clinical neurotoxicology* edited by Spencer S.S., and H. H. Schaumburg, 326-35. Baltimore, MD: Williams and Wilkins.

Crapper D.R., S. S. Krishnan, and A. J. Dalton. 1973. Brain Aluminum Distribution in Alzheimer's Disease and Experimental Neurofibrillary Degeneration. *Science* 180(4085):511-3.

Crapper DR, and AJ Dalton. 1973. Aluminum Induced Neurofibrillary Degeneration, Brain Electrical Activity and Alterations in Acquisition and Retention. *Physiology and Behavior* 10(5):935-45.

English, A., R. D. Savage, P. G. Britton, M. K. Ward, and D. N. S. Kerr. 1978. Intellectual Impairment in Chronic Renal Failure. *British Medical Journal* 1:888-90.

Exley, Christoperhor and Emily R. House. 2011. Aluminium in the Human Brain. *Monatsh Chem* 142:357-63.

Exley, Christopher. 2014. Why Industry Propaganda and Political Interference Cannot Disguise the Inevitable Role Played by Human Exposure to Aluminum in Neurodegenerative Diseases, Including Alzheimer's Disease. *Frontiers in Neurology* 5(212). doi:10.3389/fneur.2014.00212.

Garruto, Ralph M., Ryo Fukatsu, Richard Yanagihara, D. Carleton Gajdusek, Gregory Hook, and Charles E. Fiori. 1984. Imaging of Calcium and Aluminum in Neurofibrillary Tangle-bearing Neurons in Parkinsonism-dementia of Guam. *Proceedings of National Academy of Sciences of the United States of America* 81:1875-9.

Gherardi, Romain Kroum, Housam Eidi, Guillemette Crépeaux, Franãsis Jerome Authier, and Josette Cadusseau. 2015. Biopersistence and Brain Translocation of Aluminum Adjuvants of Vaccines. *Frontiers in Neurology* 6(4). doi:10.3389/fneur.2015.00004.

Ghetti, B., M. Musicco, J. Norton, and O. Bugiani.1985. Nerve Cell Loss in the Progressive Encephalopathy Induced by Aluminum Powder. A Morphologic and Semiquantitative Study of The Purkinje Cells. *Neuropathology and Applied Neurobiology* 11(1):31-53.

Golub, MS, and RP Tarara. 1999. Morphometric Studies of Myelination in the Spinal Cord of Mice Exposed Developmentally to Aluminum. *Neurotoxicology* 20(6):959-9.

Gupta, Veer Bala, S. Anitha, M. L. Hegde, L. Zecca, R. M. Garruto, R. Ravid, S. K. Shankar, R. Stein, P. Shanmugavelu, and K. S. Jagannatha Rao. 2005. Aluminium in Alzheimer?s Disease: Are We Still at a Crossroad? *Cellular and Molecular Life Sciences* 62(2):143-58.

Hammond, D.C. 2006. Quantitative Electroencephalography Patterns Associated with Medical Conditions. *Biofeedback* 34:87-94.

Han, Sungwon, Joseph Lemire, Varun P. Appanna, Christopher Auger, Zachary Castonguay, and Vasu D. Appanna. 2013. How Aluminum, an Intracellular ROS Generator Promotes Hepatic and Neurological Diseases: The Metabolic Tale. *Cell Biology and Toxicology* 29(2):75-84.

Iregren, A., B. Sjögren, K. Gustafsson, M. Hagman, L. Nylen, W. Frech, M. Andersson, K. Ljunggren, and A. Wennberg. 2001. Effects on the Nervous System in Different Groups of Workers Exposed to Aluminium. *Occupational and Environmental Medicine*, 48:453-60.

Kawahara, Masahiro, and Midori Kato-Negishi. 2011. Link between Aluminum and the Pathogenesis of Alzheimer's Disease: The Integration of the Aluminum and Amyloid Cascade Hypotheses. *International Journal of Alzheimer's Disease* doi:10.4061/2011/276393.

Kawahara, Masahiro, Midori Kato, and Yoichiro Kuroda. 2001. Effects of Aluminum on the Neurotoxicity of Primary Cultured Neurons and on the Aggregation of β-amyloid Protein. *Brain Research Bulletin* 55(2):211-17.

Lapresle, J., S. Duckett, P. Galle, and L. Cartier. 1975. Clinical, Anatomical and Biophysical Data on a Case of Encephalopathy with Aluminum Deposition. *Comptes Rendus Des Seances De La Societe De Biologie Et De Ses Filiales* 169(2):282-5. [PubMed Abstract].

Liang, Rui-Feng, Wei-Qing Li, Xiao-Hui Wang, Hui-Fang Zhang, Hong Wang, Jun-Xia Wang, Yu Zhang, Ming-Tao Wan, Bao-Long Pan, and Qiao Niu. 2012. Aluminium-Maltolate-induced Impairment of Learning, Memory and Hippocampal Long-term Potentiation in Rats. *Industrial Health* 50(5):428-36.

Lidsky, Theodore, I. 2014. Ist the Aluminum Hypothesis Dead? *Journal of Occupational and Environmental Medicine* 56(5S):S73-S79.

Lukiw, Walter J. 2010. Evidence Supporting a Biological Role for Aluminum in Chromatin Compaction and Epigenetics. *Journal of Inorganic Biochemistry* 104(9):1010-2.

Malaki, Majid. 2013. Acute Encephalopathy following the Use of Aluminum Hydroxide in a Boy Affected with Chronic Kidney Disease. *Journal of Pediatric Neurosciences* 8(1): 81-2.

Manyam B.V. 2005. What is and What is not 'Fahr's disease'. *Parkinsonism and Related Disorders* 11(2):73-80.

Markesbery, William R., William D. Ehmann, Mohammad Alauddin, and T. I. M. Hossain. 1984. Brain Trace Element Concentrations in Aging. *Neurobiology of Aging* 5:19–28.

McLaughlin, A. I. G., G. Kazantzis, E. King, D. Teare, R. J. Porter, and R. Owen. 1962. Pulmonary Fibrosis and Encephalopathy Associated with the Inhalation of Aluminium Dust. *British Journal of Industrial Medicine* 19:253-63.

Metwally, Fateheya, and Manal Mazhar. 2007. Effect of Aluminium on the Levels of Some Essential Elements in Occupationally Exposed Workers. *Archives of Industrial Hygiene and Toxicology* 58(3):305-11.

Miu, Andrei, and Oana Benga. 2006. Aluminum and Alzheimer's Disease: A New Look. *Journal of Alzheimer's Disease* 10(2-3):179-201.

Moussa Fatma, Moustafa Akmal, Zyada Fadia, and Dalia Abd Ei Hamid. 2013. Psychiatric Evaluation of a Group of Workers in the Aluminium Industry. *Egyptian Journal of Psychiatry* 34:1-9.

Mujika, J. I., F. Ruipérez, I. Infante, J. M. Ugalde, C. Exley, and X. Lopez. 2011. Pro-oxidant Activity of Aluminum: Stabilization of the Aluminum Superoxide Radical Ion. *The Journal of Physical Chemistry A* 115(24):6717-23.

Nayak, Prasunpriya and Ajay K. Chatterjee. 2001. Effects of Aluminum Exposure on Brain Glutamate and GABA Systems. *Food and Chemical Toxicology* 39(5):1285-9.

Nayak, Prasunpriya, S. B. Sharma, and N. V. S. Chowdary. 2010. Augmentation of Aluminum-Induced Oxidative Stress in Rat Cerebrum by Presence of Pro-oxidant (Graded Doses of Ethanol) Exposure. *Neurochemical Research* 35(11):1681-90.

Nayak, Prasunpriya, S. B. Sharma, and N. V. S. Chowdary. 2011. Impact of Coexposure to Aluminum and Ethanol on Phosphoesterases and Transaminases of Rat Cerebrum. *Journal of Medical Biochemistry* 30:1-5.

Nayak, Prasunpriya, S. B. Sharma, and N. V. S. Chowdary. 2012. Pro-oxidant Status Based Alterations in Cerebellar Antioxidant Response to Aluminum Insult. *Neurochemical Journal* 6(1):44-52.

Nayak, Prasunpriya, S. B. Sharma, and N. V. S. Chowdary. 2013. Aluminum and Ethanol Induce Alterations in Superoxide and Peroxide Handling Capacity (SPHC) in Frontal and Temporal Cortex. *Indian Journal of Biochemistry and Biophysics* 50(5):402-10.

Nayak, Prasunpriya, S. B. Sharma, and N. V. S. Chowdary. 2014. Oxidant Handling by Hippocampus and Hebb-William Maze Performance in Aluminum-exposed Albino Wistar Rats. *International Journal of Clinical and Experimental Physiology* 1(2):106-12.

Nayak, Prasunpriya, S. B. Sharma, and N. V. S. Chowdary. 2015. Influence of Prooxidant Exposure on Aluminium-induced Alterations in Open Field Behavioral Study in Rats. *Indian Journal of Physiology and Pharmacology* 59(3):298-307.

Nayak, Prasunpriya. 2006. Perinatal Toxicity of Aluminum. *The Internet Journal of Toxicology* 3(1). Accessed August 8, 2015.

Nayak, Prasunpriya. 2012. Conjecturable Role of Aluminum in Pathophysiology of Stroke. In *Metal Ion in Stroke, Springer Series in Translational Stroke Research* edited by Li, Y.V., and J. H. Zhang. 649-80. Springer Science+Business Media, LLC.

Nayak, Prasunpriya. 2014. Prooxidative and Antioxidative Intervention of Aluminum-Induced Oxidative Stress and Neurodegeneration. PhD diss., Chettinad Academy of Research and Education.

Platt, Bettina, Graham Fiddler, Gernot Riedel, and Zaineb Henderson. 2001. Aluminium Toxicity in the Rat Brain: Histochemical and Immunocytochemical Evidence. *Brain Research Bulletin* 55(2):257-67.

Rajput, A.H., Ryan J. Utti, B. Rozdilsky. 1985. Distribution of Metals in Parkinson's Disease and Control Brains. *Neurology* 35(Suppl 1):224.

Riihimäki, Vesa, Helena Hänninen, R. Akila, T. Kovala, Eeva Kuosma, H. Paakkulainen, Sinikka Valkonen, and Berndt Engström. 2000. Body Burden of Aluminum in Relation to Central Nervous System Function among Metal Inert-gas Welders. *Scandinavian Journal of Work, Environment and Health* 26(2):118-30.

Sánchez-Iglesias, Sofía, Estefanía Méndez-Álvarez, Javier Iglesias-González, Ana Muñoz-Patiño, Inés Sánchez-Sellero, José Luís Labandeira-García, and Ramón Soto-Otero. 2009. Brain Oxidative Stress and Selective Behaviour of Aluminium in Specific Areas of Rat Brain: Potential Effects in a 6-OHDA-induced Model of Parkinson's Disease. *Journal of Neurochemistry* 109(3):879-88.

Sarkaki, A., S. Zahedi Asl, and R. Assaei. 2009. Effect of Intrahippocampal Injection of Aluminum on Active Avoidance Learning in Adult Male Rats. *Pakistan Journal of Biological Sciences* 12(1):40-5.

Savory J., M. M. Herman, C. D. Katsetos, and M. R. Wills. 1991. Aluminum and neurodegenerative disorders. In *Aluminum in chemistry, biology and medicine* edited by Nicolini M., P. F. Zatta, and B. Corain. 45-52. Cortina International, Verona.

Selkoe, Dennis J., Ronald K. H. Liem, Shu-Hui Yen, Michael Shelanski. 1979. Biochemical and Immunological Characterization of Neurofilaments in Experimental Neurofibrillary Degeneration. *Brain Research* 163(2):235-52.

Seneff, Stephanie, Robert Davidson, and Jingjing Liu. 2012. Empirical Data Confirm Autism Symptoms Related to Aluminum and Acetaminophen Exposure. *Entropy* 14(11):2227-53.

Shaw Chirostopher A. and Michael S. Petrik. 2009. Aluminum Hydroxide Injections Lead to Motor Deficits and Motor Neuron Degeneration. *Journal of Inorganic Biochemistry* 103(11):1555-62.

Shaw, C. A., S. D. Kette, R. M. Davidson, and S. Seneff. 2013. Aluminum's Role in CNS-immune System Interactions Leading to Neurological Disorders. *Immunome Research* 9 069. doi: 10.4172/1745-7580.1000069.

Shaw, Christopher A. and Thomas E. Marler. 2013. Aluminum and the Human Diet Revisited. *Communicative and Integrative Biology* 6 doi:10.4161/cib.26369.

Shaw, Christopher A., Stephanie Seneff, Stephen D. Kette, Lucija Tomljenovic, John W. Oller Jr., and Robert M. Davidson. 2014. Aluminum-Induced Entropy in Biological Systems: Implications for Neurological Disease. *Journal of Toxicology* doi:10.1155/2014/491316.

Shi-Lei, Sun, Ma Guang-Yu, Lihua Bachelor, Zhuyun Bachelor, Dong Hong-Mei, and Xu Xiao-Hu. 2005. Effect of Naloxone on Aluminum-induced Learning and Memory Impairment in Rats. *Neurology India* 53(1):79-82.

Sinczuk-Walczak, Halina, Maria Szymczak, Grazyna Razniewska, Wandda Matczak and Wieslaw Szymczak. 2003. Effects of Occupational Exposure to Aluminum on Nervous System: Clinical and Electroencephalographic Findings. *International Journal of Occupational Medicine and Environmental Health* 16(4):301-10.

Sjogren Bengt, Anders Iregren, Wolfgang Frech, Maud Hagman, Lotta Johansson, Maria Tesarz, and Arne Wennberg. 1996. Effects on the Nervous System among Welders Exposed to Aluminium and Manganese. *Occupational and Environmental Medicine* 53:32-40.

Skalasky, Harry, and Richard Carchman. 1983. Aluminum Homeostasis in Man. *International Journal of Toxicology* 2(6):405-23.

Speziali, Margherita, and Edoardo Orvini. 2003. Metals Distribution and Regionalization in the Brain. In *Metal Ions and Neurodegenerative Disorders* edited by Zatta Paolo. 15-66. World Scientific Publishing Co. Pte. Ltd. Singapore.

Terry, Robert D., and Carlos Peña. 1965. Experimental Production of Neurofibrillary Degeneration. *Journal of Neuropathology and Experimental Neurology* 24(2):200-10.

Thippeswamy, Agadi Hiremath, Mohamed Rafiq, Gollapalle Lakshminarayana shastry Viswantha, Kethaganahalli J. Kavya, Suryakanth D. Anturlikar, and Pralhad S. Patki. 2013. Evaluation of Bacopa monniera for its Synergistic Activity with Rivastigmine in Reversing Aluminum-Induced Memory Loss and Learning Deficit in Rats. *Journal of Acupuncture and Meridian Studies* 6(4):208-13.

Tomljenovic, Lucija. 2011. Aluminum and Alzheimer's Disease: After a Century of Controversy, Is There a Plausible Link? *Journal of Alzheimer's Disease* 23(4):567-98.

Traub, R. D., T. C. Rains, R. M. Garruto, D. C. Gajdusek, and C. J. Gibbs. 1981. Brain Destruction Alone Does Not Elevate Brain Aluminum. *Neurology* 31(8):986.

Tripathi, Sandeep, Abbas Ali Mahdi, Akbar Nawab, Ramesh Chander, Mahdi Hasan, Mohammad Shakil Siddiqui, Farzana Mahdi, Kalyan Mitra, and Virendra Kumar Bajpai. 2009. Influence of Age on Aluminum Induced Lipid Peroxidation and Neurolipofuscin in Frontal Cortex of Rat Brain: A Behavioral, Biochemical and Ultrastructural Study. *Brain Research* 1253:107-16.

Troncoso, Juan C., Donald L. Price, John W. Griffin, and Irma M. Parhad. 1982. Neurofibrillary Axonal Pathology in Aluminum Intoxication. *Annals of Neurology* 12(3):278-83.

Utti, Ryan J., A. H. Rajput, B. Rozdilsky, M. Bickis, T. Wollin and W. K. Yuen. 1989. Regional Metal Concentrations in Parkinson's Disease, Other Chronic Neurological Diseases, and Control Brains. *Canadian Journal of Neurological Sciences* 16:310-4.

Uversky, Vladimir N., Jie Li, and Anthony L. Fink. 2001. Metal-triggered Structural Transformations, Aggregation, and Fibrillation of Human α-Synuclein: A Possible Molecular Link Between Parkinson'S Disease and Heavy Metal Exposure. *Journal of Biological Chemistry* 276:44284-96.

Varner, Julie A, Karl F Jensen, William Horvath, and Robert L Isaacson. 1998. Chronic Administration of Aluminum–fluoride or Sodium–fluoride to Rats in Drinking Water: Alterations in Neuronal and Cerebrovascular Integrity. *Brain Research* 784(1-2):284-98.

Winter, Arthur, and Winter Ruth. 2007. *Smart Food: Diet and Nutrition for Maximum Brain Power.* 2021 Pine Lake Road, Suite 100, Lincoln, NE 68512. ASJA Press. iUniverse.

Yasui, M., T. Kihira, K. Ota. 1992. Calcium, Magensium and Aluminum Concentrations in Parkinson's Disease. *Neurotoxicology* 13:593-600.

Yegambaram, Manivannan, Bhagyashree Manivannan, Thomas G. Beach, and Rolf U. Halden. 2015. Role of Environmental Contaminants in the Etiology of Alzheimer's Disease: A Review. *Current Alzheimer Research* 12:116-46.

Yeh, Yen-Hung, Ya-Ting Lee, Hung-Sheng Hsieh, and Deng-Fwu Hwang. 2009. Effect of Taurine on Toxicity of Aluminum in Rats. *E-SPEN The European E-Journal of Clinical Nutrition and Metabolism* 4(4):E187-192. doi:10.1016/j.eclnm.2009.05.013.

Yokel Robert A. 2013. Aluminum. In *Encyclopedia of Human Nutrition* edited by Caballero B. 57-63. Waltham, MA: Academic Press.

Yokel, Robert A. 1994. Aluminum Chelation: Chemistry, Clinical, and Experimental Studies and the Search for Alternatives to Desferrioxamine. *Journal of Toxicology and Environmental Health* 41(2):131-74.

Yu, Lijuan., Rong Jiang, Qiang Su, Huarong Yu, and Junqing Yang. 2014. Hippocampal Neuronal Metal Ion Imbalance Related Oxidative Stress in a Rat Model of Chronic Aluminum Exposure and Neuroprotection of Meloxicam. *Behavioral and Brain Functions* 10:6. doi:10.1186/1744-9081-10-6.

Yumoto, Sakae, Hisao Nagai, Hiroyuki Matsuzaki, Hiroshi Matsumura, Wataru Tada, Emiko Nagatsuma, and Kouichi Kobayashi. 2001. Aluminium Incorporation into the Brain of Rat Fetuses and Sucklings. *Brain Research Bulletin* 55(2):229-34.

Zatta, Paolo F., Nicolini M, and Corain, B. 1991. Aluminum (III) Toxicity and Blood-Brain Barrier Permeability. In *Aluminum in Chemistry, Biology and Medicine* edited by Nicolini, M., Zatta P. F., and Corain B. 97-112. Cortina International. Verona.

Zhang, QL, P. Boscolo, PY Niu, F. Wang, YT Shi, L. Zhang, LP Wang, J. Wang, M. Di Gioacchino, P. Conti, QY Li, and Q. Niu. 2008. How Do Rat Cortical Cells Cultured with Aluminum Die: Necrosis or Apoptosis? *International Journal of Immunopathology and Pharmacology* 21(1):107-15.

In: Aluminum Neurotoxicity
Editor: Ankica Jelenković
ISBN: 978-1-63484-739-1

Chapter 6

LINK BETWEEN ALUMINUM NEUROTOXICITY AND NEURODEGENERATIVE DISORDERS WITH A FOCUS ON BETA-AMYLOID

Masahiro Kawahara
Department of Bio-Analytical Chemistry,
Faculty of Pharmacy, Musashino University, Nishitokyo-shi, Tokyo, Japan

ABSTRACT

Aluminum is an old element that has been known for a long time, but some of its properties are only now being discovered. Although environmentally abundant, aluminum is not essential for life. In fact, because of its specific chemical properties, aluminum inhibits more than 200 biologically important functions and exerts various adverse effects in plants, animals, and humans. Aluminum is a widely recognized neurotoxin. It has been suggested that there is a relationship between exposure to aluminum and neurodegenerative diseases, including dialysis encephalopathy, amyotrophic lateral sclerosis and Parkinson's dementia in the Kii Peninsula and Guam, as well as Alzheimer's disease: however, this claim remains to be verified. In this chapter, we review the detailed characteristics of aluminum neurotoxicity and the link between Alzheimer's disease and other neurodegenerative diseases, based on recent findings on amyloid toxicity, metal-metal interactions, and the functions of metalloproteins in synapses.

Keywords: apoptosis, synapse, iron homeostasis, calcium homeostasis, zinc

INTRODUCTION

Aluminum (Al) is the third most abundant element in the earth's crust and is widely distributed throughout our environment. Compounds containing Al and oxygen or silicon (e.g., clay, glass, and alum) have been used as materials for centuries. Thus, Al is considered to be an old, well-known metal. However, Al was first isolated as an element in 1827, and its use as a silvery metal began only after 1886. From this point of view, Al is a new metal, and therefore, little is known about the effects of Al on our bodies. Because Al is light, nonmagnetic, malleable, and ductile, it has a widespread and important use in industrial applications and in manufacturing of consumer products. Al is also used to produce cooking utensils and pharmacological agents such as antacids and antiperspirants, through which it can enters the human body. Moreover, Al compounds, such as aluminum sulfate or aluminum hydroxyl polymer, are widely used in the process of purifying drinking water.

Despite its widespread distribution throughout the environment, Al is not essential for life beings. On the contrary, Al is a well-established neurotoxin. An association between Al poisoning and memory disorder in humans was first reported in 1921 [1]. Later, the intracerebral administration of Al reportedly induced epilepsy in experimental animals [2]. As a component of dialysis solution or some pharmacological compounds, Al is known to cause a variety of dialysis-related disorders, including osteomalacia (aluminum bone disease), microcytic anemia, β_2-microglobulin-associated amyloidosis, and dialysis encephalopathy in hemodialysis patients [3, 4]. An accidental contamination of drinking water by Al occurred in Camelford, Cornwall in the UK, in 1988, exposing more than 20,000 persons to high levels of Al [5]. These individuals exhibited various symptoms related to cerebral impairment, such as loss of concentration and short term memory, in a 10-year follow-up study. Furthermore, Al is suspected to be linked to various neurodegenerative diseases including Alzheimer's disease (AD) [6-10], amyotrophic lateral sclerosis (ALS) and Parkinson's disease (PD) in the Kii Peninsula and Guam [11]. Associations of Al with Gulf-war syndrome and autism are also suspected, since Al is used as an adjuvant in vaccines [12, 13]. In particular, the link between Al and AD has been the subject of much debate in the past few decades [14]. Numerous reviews have discussed the association between Al and neurodegenerative diseases; however, in this review we focus on the interaction of Al and other metals with other disease-related proteins, based on recent findings, and revisit the significance of the Al in the pathogenesis of AD and other neurodegenerative diseases.

Effects of Aluminum on the Central Nervous System and Its Chemical Characteristics

Chemical Characteristics of Aluminum

Al is the third most abundant element in the Earth's crust. The amount of Al in the Earth's crust —its lithospheric abundance— is 82000 ppm. Meanwhile, the amount of Al in the human body —the biospheric abundance — is only 0.9 ppm. Indeed, the B/L (biospheric/lithospheric) ratio of Al (approximately 1.1×10^{-5}) is extremely low compared to other essential elements such as Ca (0.35), Fe(1.5×10^{-3}), Zn(0.44) [15]. Therefore, despite its abundancy, Al is not essential for life. This is due to its specific chemical characteristics [16].

Al exhibits only one oxidation state, Al^{3+}. In acidic solutions with pH<4, Al^{3+} exists as a soluble octahedral hexahydrate $Al(H_2O)_6^{3+}$ (Al^{3+}). As the pH increase, its solubility decreases and $Al(OH)_3$ precipitates at neutral pH. Thus, the concentration of free Al^{3+} in the physiological condition is usually very low. This chemistry suggest a low level of Al were present in the seawater when living organisms appeared.

Al^{3+} has affinity for negatively charged, oxygen-donor ligands. Inorganic and organic phosphates, carboxylate, and deprotonated hydroxyl groups form strong bonds with Al^{3+}. Owing to these chemical characteristics, Al^{3+} binds to the phosphate groups of DNA and RNA, affecting DNA topology and influencing the expression of various genes essential for brain functions. Al^{3+} also binds to the phosphate groups of nucleoside di- and triphosphates, such as adenosine triphosphate (ATP) and can thus influence energy metabolism. Al also inhibits the functions of various protein kinases and phosphatases.

Al^{3+} has very low ligand-exchange rate in comparison to other metals. For example, the ligand-exchange rate of Mg^{2+} is 10^5 times faster than that of Al^{3+}, and therefore, Al^{3+} inhibits enzymes with Mg^{2+} cofactors. Al^{3+} also inhibits biological processes involving rapid Ca^{2+} exchange: the exchange rate for Al^{3+} is 10^8 times slower than that of Ca^{2+}. Due to these properties, Al cannot participate in Ca^{2+}- or Mg^{2+} -related enzymatic reactions, and has an expanded half-life in the body. Moreover, Al^{3+} has the characteristics similar to those of iron (Fe), and binds to Fe-binding proteins such as transferrin and ferritin.

Al^{3+} has strong positive charges and a relatively small ionic radius in comparison to other metal ions such as Ca^{2+}, Zn^{2+}, and Na^+ (Figure 1). Thus, Al^{3+} firmly binds to metal-binding amino acids (histidine [His], tyrosine [Tyr], arginine [Arg] etc.) or phosphorylated amino acids. Al is a well-known cross-linker of proteins and is therefore used as a leather tanning agent.

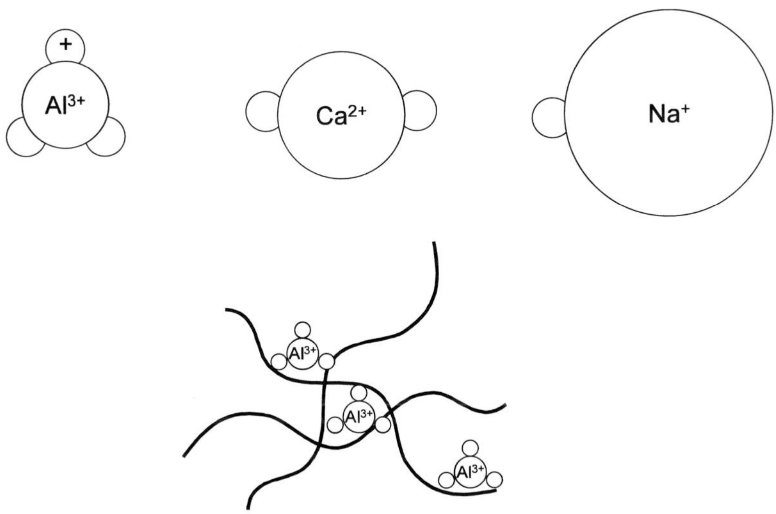

Figure 1. Cross linking of protein by Al^{3+}.
Al^{3+} has a relatively small ionic radius (50 pm) with three positive charges; here it is compared to other metal ions such as Zn^{2+} (74 pm), Ca^{2+} (99 pm), and Na^+ (95 pm). These characteristics enable Al to be an effective cross-linker of proteins.

Effects of Aluminum on the Central Nervous System

Owing to these chemical characteristics, Al is not an essential element and no enzymatic reaction requires Al. In contrast, Al has been reported to influence more than 200 biologically important reactions and to cause various adverse effects on the mammalian central nervous system (CNS) as shown in Table 1 [10]. These include processes crucial for brain functions, such as axonal transport, neurotransmitter synthesis, synaptic transmission, phosphorylation or dephosphorylation of proteins, protein degradation, gene expression, and inflammatory responses.

Table 1. Effects of aluminum on the central nervous system

1. Nucleus and gene expression
Binding to DNA
binds to histone-DNA complex and induces conformational changes of chromatin
induces topological changes of DNA
Altered gene expression
induces decreased expression of neurofilament and tubulin
induces altered expression of genes of neurofilament, APP, and neuron specific enolase
induces decreased expression of transferrin receptor
induces altered expression of RNA polymerase I
induces down regulation of mitochondrial cytochrome c oxidase
induces altered expression of calbindin-D28k
induces decrease in the expression of nerve growth factor (NGF) and brain derived neurotrophic factor (BDNF)
induces expression of pro-inflammatory genes and pro-apoptotic genes
induces elevated expression of APP
induces altered expression of oxidative stress marker genes (SOD1, glutathione reductase, etc.)
induces decreased expression of neprilysin
induces altered expression of ß-APP secretase (BACE1 and BACE2)
enhances expression of APP
2. Cellular functions
Energy metabolism
inhibits the activity of hexokinase
inhibits the activity of phosphofructokinase
inhibits the activity of glucose-6-phosphate dehydrogenase
causes mitochondorial dysfunction and depletion of ATP
decreases in activity and expression of TCA-cycle related enzymes (succinate dehydrogenase (SDH), alpha-ketoglutarate dehydrogenase (KGDH), isocitrate dehydrogenase-NAD^+ (IDH), fumarase (FUM), aconitase (ACN), and cytochrome c oxidase (Cyt C Ox)).
Phosphorylation and dephosphorylation
inhibits the activity of protein phosphatase
increases the activity of protein kinase C and cytoskeleton proteins
accelerates phosphorylation and accumulation of neurofilament
enhances Ca^{2+}/Calmodulin dependent protein kinase activity
accelerates phosphorylation of MAP 2 and neurofilament
inhibits dephosphorylation of tau
induces non-enzymatic phosphorylation of tau
Conformational changes and abnormal accumulation of proteins
causes the conformational change and the accumulation of neurofilament and MAP1A, MAP1B
accelerates the phosphorylation of tau and its accumulation
causes the accumulation of tau protein in neuroblastoma cells or in primary cultured neurons
causes the accumulation of tau protein in experimental animals

Conformational changes and abnormal accumulation of proteins
causes neurofibrillary degeneration *in vivo*
causes the accumulation of AßP in cultured neurons or in neuroblastoma cells
causes the accumulation of AßP *in vivo*
causes the conformational change of human islet amyloid protein (amylin)
causes the conformational change of ABri (Familial British dementia-related protein)
causes the conformational change of Ataxin 3
causes the conformational change of $ß_2$-microglobulin
Neurotransmitter release
inhibits glutamate release
impairs synaptic transmission
inactivates glutamate dehydrogenase
inhibits NMDA-type glutamate receptor
inhibits choline acetyl transferase and tyrosine hydroxylase, glutamate decarboxylase,
influences acetyl-CoA and inhibits acetylcholine release
activates monoamine oxidase
inhibits dopamine beta-hydroxylase
inhibits uptake of serotonin and noradrenalin in synaptosomes
Channel inhibition
influences the activities of Na^+ channels and K^+ channels
enhances the voltage-activated Na^+ channels
inhibits the voltage-gated calcium channel
inhibits the IP_3-mediated Ca^{2+} release
Others
influences GTP binding proteins as aluminum fluoride
inhibits GAP junction
inhibits axonal transports
binds to calmodulin and inhibition of calmodulin-binding enzymes
induces inflammetry responses
3. Membrane lipids
Peroxidation
accelerates iron-induced membrane lipid peroxidation
enhances lipid peroxidation in liposomes
induces peroxidation of myelin lipids *in vivo*
increases peroxidation products (malondialdehyde)
Membrane properties
causes the change the lipid/phospholipids profiles of myelin *in vivo*
induces the change in membrane physical properties (surface potential, lipid fluidity, and lipid arrangement)
induces the change of membrane fluidity
4. Higher functions
Cell death
causes the apoptotic neuronal death
causes the apoptosis of astrocytes
causes the death of motor neuron
Behavior, learning, and memory, others
inhibits long term potentiation (LTP)
causes learning disorder or memory deficit in experimental animals
influences electrical activity in hippocampus and inhibits spatial learning memory deficit in aging rats
causes memory deficit in AD model mice
causes encephalopathy in dialysis patients
causes encephalopathy in patients with renal failure

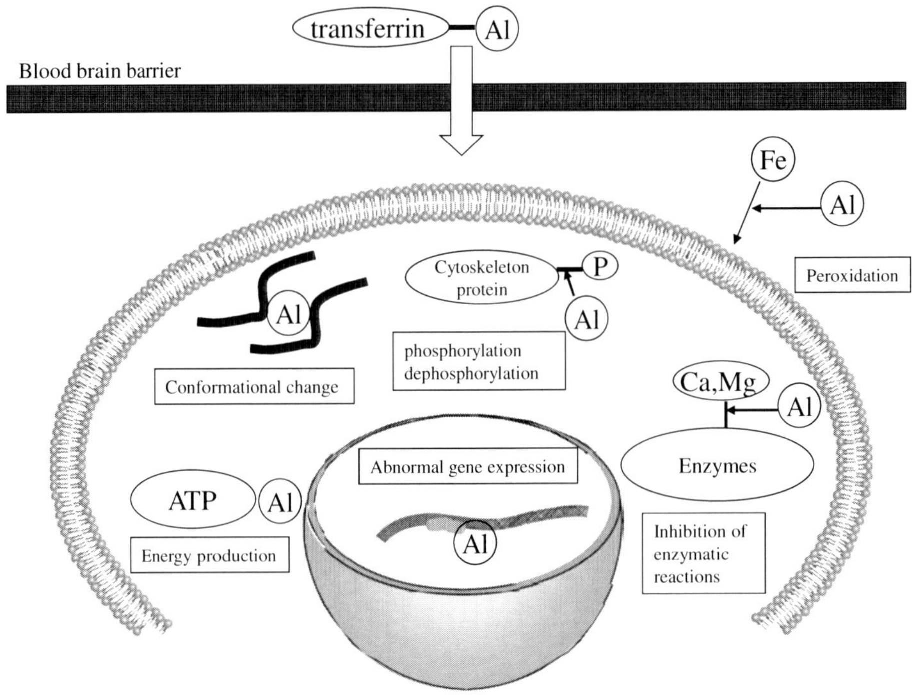

Figure 2. Effects of aluminum on the central nervous system.
Major effects of Al on the central nervous system are depicted.

By binding to transferrin, Al can cross blood-brain barrier and enter into the brain. Al influences the expression of various genes crucial for brain functions. Lukiw et al. reported that nanomolar levels of Al^{3+} were sufficient to influence neuronal gene expression [17, 18]. Al inhibits numerous Ca^{2+}-related or Mg^{2+}-related enzymes, and protein kinases or phosphatases [19]. Strong binding of Al^{3+} to phosphorylated amino acids promotes the self-aggregation and accumulation of highly phosphorylated cytoskeletal proteins, including neurofilament and microtubule-associated proteins (MAPs) etc. [20]. By binding to various proteins including Alzheimer's ß-amyloid protein (A ßP), Al causes the oligomerization of proteins, inducing conformational changes that can inhibit their degradation by proteases. Conformational changes of these proteins induced by Al causes various adverse effects and triggers the pathogenesis of neurodegenerative diseases.

Consequently, Al causes apoptotic death of neurons and glial cells [21, 22]. Chronic administration of Al impairs long-term potentiation (LTP), which is a form of synaptic information storage well-known as a paradigm of memory mechanisms [23]. Al also impairs various enzymes including those related to neurotransmitter synthesis and thus affects the neurotransmitter content. Al^{3+} also inhibits voltage-gated Ca^{2+} channels and neurotransmitter receptors, and impairs synaptic transmission [24]. Finally, Al causes spatial memory deficit, influences emotional reactivity, and impairs various brain functions related to learning and memory [25, 26]. These adverse effects may be involved in the mechanisms that underlie Al-induced memory disorder. Figure 2 depicts the typical effects of Al in the central nervous system.

LINK BETWEEN ALUMINIUM AND ALZHEIMER'S DISEASE

Aluminum Hypothesis: Historical Overview

Among various neurodegenerative diseases, the association between Al and AD has been of particularly interest. A possible relationship between Al and the pathogenesis of AD has been discussed for several decades. AD is a severe senile type of dementia that affects a large section of the elderly population worldwide. In Japan, there were more than 4,000,000 patients with senile dementia at 2013, and the number is increasing annually. The number of patients with mild cognitive incidence (MCI), which is the precursor stage of senile plaque, is also estimated to be more than 4,000,000.

AD is characterized by profound memory loss and the inability to form new memories [27]. The pathological hallmarks of AD are the presence of numerous extracellular deposits, including senile plaques and neurofibrillary tangles (NFTs), and the selective loss of synapses and neurons in the hippocampal and cerebral regions. Indeed, there is a strong correlation between the decrease in the number of synapses and the severity of memory impairment [28]. The major components of NFTs and senile plaques are phosphorylated tau protein and ß-amyloid protein (AßP), respectively. AD is a multifactorial disease and genetic factors as well as environmental factors play essential roles in its pathogenesis. Al has been considered to be one such environmental risk factor, and its potential role in the pathogenesis of AD has been termed the "aluminum hypothesis", in which Al is considered to be a risk factor for, and a contributor to the pathogenesis of AD.

The link between Al and AD has been supported by many studies, beginning in 1965 with the finding of Klatzo et al. that the intracerebral administration of Al to experimental animals induced degeneration of neurofibrils and the appearance of tangle-like structures that were similar to the NFTs found in the brains of AD patients [29]. It was subsequently reported by Crapper et al. that Al levels were elevated in the brains of AD patients [30]. Furthermore, in the 1970s, the Al present in dialysis solution and in pharmacological compounds was found to cause dementia in dialysis patients (dialysis encephalopathy). Almost two decades later, Martyn et al. reported a high percentage of AD cases in areas with a high level of Al present in the drinking water [31]. Chelation therapy using desferoxamine (DFO), an Al-chelator, was effective in the maintenance of daily living skills in AD patients [32].

Despite such supporting evidence, the aluminum hypothesis of AD remains controversial and has been the subject of much debate over the past few decades. There were at least three arguments against the aluminum hypothesis [14, 33, 34]. *First*, it has been argued that the neurofibrillary changes observed in Al-intoxicated animals (Al-NFTs) differ from those in AD patients (AD-NFTs). These include morphological and biochemical differences such as the lack of paired helical filamental (PHF) structures in Al-NFTs, the different distributions of Al-NFTs and AD-NFTs at nerve terminals, and the absence of immunoreactivity for tau protein in Al-NFTs, although this is the main component of AD-NFTs. *Second*, there is no significant difference in the Al levels in AD patients and age-matched non-AD controls. *Third*, the epidemiological studies on Al in drinking water are premature and inconclusive. However, most of these criticisms were levied before the 1990s. We would like to re-investigate these early arguments in the context of new findings in the study of AD.

Aluminum-Induced Pathological Changes

Regarding the first argument, more recent immunohistochemical studies have indicated that depositions in the brains of Al-intoxicated animals do stain with the anti-tau antibody. The accumulation of tau protein has been reported in patients with dialysis encephalopathy, and in cultured neuronal cells exposed to Al [35, 36]. Al has been shown to inhibit the dephosphorylation of tau [37] and to enhance its aggregation *in vitro* [38]. Furthermore, NFTs in some AD patients have been shown to be composed of straight-type filaments rather than the PHF-type filaments as is observed in Al-NFT [39]. These data indicate that attempts to discredit the aluminum hypothesis on the basis of differences between Al-NFTs and AD-NFTs are no longer tenable.

Al-induced pathological changes were first attributed to tau proteins in NFTs early in this controversy. However, after the 1990's, it has been widely accepted that the accumulation of AßP and its neurotoxicity play a central role in the pathogenesis of AD [40]. Thus, the evidence indicating an accumulation of AßP in Al-intoxicated neurons is of importance. We have demonstrated that the chronic application of Al causes the accumulation of AßP and tau protein in cultured neurons derived from rat cerebral cortex [41]. Al has also been shown to induce an elevation in AßP levels in neuroblastoma cells [42]. Pratico et al. found that Al-fed mice that had been transfected with the human amyloid precursor protein (APP) gene (Tg 2576 mice) exhibited pathological changes similar to those seen in the human AD brain, including a marked increase in the amount of both secreted and accumulated AßP, and increased deposition of senile plaques [43]. An important aspect of these findings is that orally administered Al enhanced the accumulation of AßP in the brain of AD-model mice. Oral Al exposure has been shown to impair spatial learning and neurogenesis in Tg 2576 mice alongside an accumulation of AßP [44]. A 58-year woman who was accidentally exposed to Al at 1988 in Camelford (Cornwall, U.K.) died 15 years after the incident. The patient exhibited neurological symptoms of memory disorders, a marked deposition of Al, and a rare form of cerebral amyloid angiopathy in the brain [45]. Considering that cerebral amyloid angiopathy is characterized by the deposition of AßP in blood vessels and has a causative link with AD [46], these findings indicate that the exposure to Al, even if it is short-term, could cause the accumulation of AßP and induce adverse effects in human.

Accumulation of Aluminium in Alzheimer's Disease Brains

The second argument cites a lack of significant differences between Al levels in AD patients and age-matched non-AD controls. One reason for this controversy may be the Al contamination of the solutions used in the process of tissue fixation and staining. Therefore, prior studies in fixed tissues cannot be relied upon for precise assessment of the presence and levels of Al; quantitative analysis of non-fixed and freshly frozen tissues is necessary. One such study showed that the amount of Al in whole brains of AD patients was not significantly different from that in control brains [47]. However, in a recent study of a transgenic mouse model of AD, the accumulation of Al and AßP was demonstrated, although there was no significant difference in Fe, Zn, and Cu levels between the AD model mice and their wild type controls [48]. Furthermore, the analysis of the metal profiles in the sera of AD patients and that of patients with MCI demonstrated that the amount of Al increases significantly as dementia process [49]. Additionally, a recent meta-analyses study of Al levels in AD patients demonstrated that significantly higher levels of Al was found in the brains, sera, and CSF of AD patients from 34 studies [50]. Arain et al. demonstrated that levels of Al and Mn in scalp hair in patients of AD and other neurodegenerative diseases are significantly higher than control [51].

The aluminum hypothesis debate has further been fueled by a failure to detect Al in the senile plaques or NFTs [52]. However, it has been argued that this failure could simply be because of the limits of detection of the analytical method used (nuclear microscopy). An accumulation of Al in both senile plaques and NFTs has been reported for renal failure patients [53]. Furthermore, Bouras et al. have reported an accumulation of Al in NFT-bearing neurons in the AD brains, after using highly sensitive laser microprobe mass analysis (LAMMA) on non-fixed brain samples [54]. Yumoto et al. analyzed Al in senile plaques of AD patients using energy-dispersive X-ray spectroscopy combined with transmission electron microscopy (TEM-EDX), a method that yields a high-resolution and low detection limit [55]. Their detailed analysis demonstrated that Al was present in the cores of senile plaques at a concentration of 35-50 ppm.

Epidemiological Studies of Aluminium in Alzheimer's disease

After the initial finding about the association between AD and Al in drinking water in 1989, a considerable number of epidemiological studies have provided evidence supporting the association between Al and AD [56], although some epidemiological studies have failed to demonstrate such a relationship [57, 58]. However, there are a number of possible explanations for this inconsistency, particularly when considering the difficulty of making side-by-side comparisons in epidemiological studies of Al (e.g., intake estimations, effect of movement, differences in water-treatment processes, etc.). Using strict neuropathological criteria to discriminate between patients with AD and controls (including histopathological verification), McLachlan et al. found that an elevated risk of histopathologically verified AD was associated with the consumption of higher concentrations of Al *via* drinking water [59]. More detailed analysis revealed an association between exposure to organic monomeric Al and AD, even after adjustment for education level, family history and the presence of the apoE4 allele [60].

The amount of Al consumed in drinking water is approximately 5% of the total daily intake. Thus, it is possible that some factors that prevent or accelerate Al absorption may exist in drinking water. Silicate in the water has been reported to interact with Al and thereby prevent Al toxicity to fish [61] or inhibit Al absorption [62]. Thus, the level of silicate in drinking water is also important. The relationship between Al and the cognitive impairment is suggested to be influenced by the silicate concentration as well as the pH in the water [63]. A 15-year follow-up in a French cohort study demonstrated that higher daily intake of Al *via* drinking water is associated with an increased risk of dementia, and that increased silicate intake is associated with a reduced risk of dementia [64]. Furthermore, the cognitive impairment among women was correlated with low concentrations of silicate in drinking water [65]. Additionally, the recent epidemiological study of workers who had been used Al suggested that occupational exposure to Al is associated with cognitive impairment [66].

In considering these new lines of evidence, it is difficult to agree with the early criticisms of the aluminum hypothesis. Thus, the significance of the Al in the pathogenesis of AD needs to be revisited. In this context, it is important to consider Al-induced conformational changes of AßP and the neurotoxicity of this protein.

Amyloid Hypothesis and Alzheimer's Disease

Numerous biochemical, cell biological, and genetic studies have supported the idea termed "amyloid cascade hypothesis", which holds the AßP accumulation and the consequent neurodegeneration play a central role in AD [40], although the precise cause of AD remains under investigation. Moreover, recent studies on the identified AßP species have indicated that the oligomerization of AßP and its conformational changes are critical in the neurodegeneration process [67, 68].

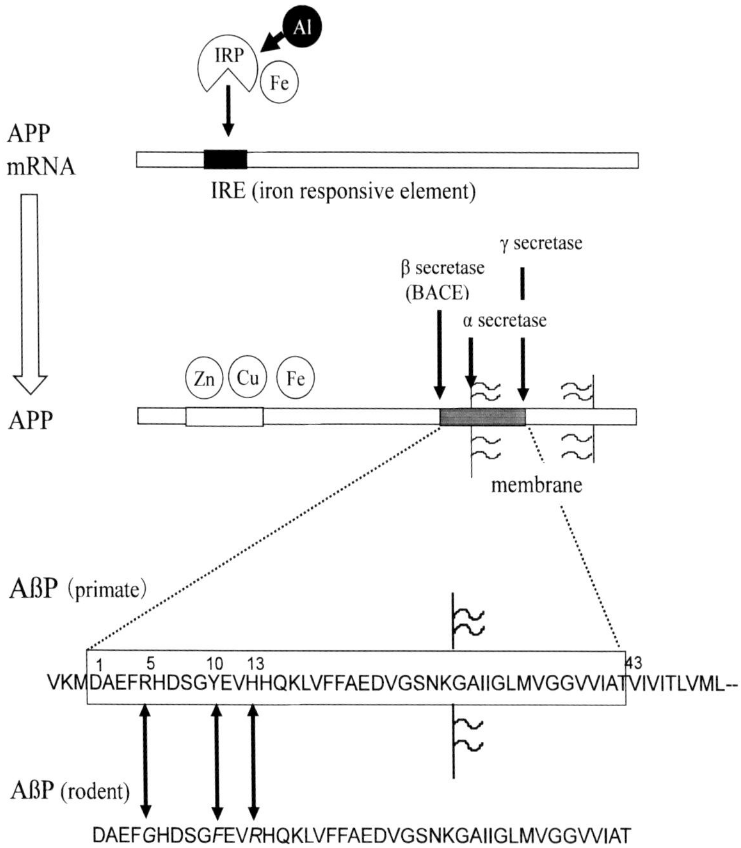

Figure 3. Secretion of AßP from APP.
AßP is secreted by the cleavage of the APP N-terminus by ß-secretase (BACE), followed by the intra-membrane cleavage of the C-terminus by γ-secretase. APP also binds to Cu or Zn. Human AßP and rodent AßP differ in three amino acids (Arg^{5}, Tyr^{10}, and His^{13}). The expression of APP is regulated by Fe and Al.

AßP is a small peptide of 39–43 amino acid residues, secreted after cleavage of the amyloid precursor protein (APP) N-terminus by the ß-APP cleaving enzyme (BACE) and intramembrane cleavage of its C-terminus by γ-secretase (Figure 3). Genetic studies of early-onset cases of familial AD have indicated that APP mutations and AßP metabolism are associated with AD [69]. Moreover, the mutations in the presenilin genes also account for the majority of cases of early-onset familial AD [70]. Presenilins are one of γ -secretases and mutations in the corresponding genes influence the production of AßP species [71].

Yankner et al. reported that the first 40 amino acid residues of AßP (AßP (1–40)) causes the death of cultured rat hippocampal neurons or neurodegeneration in the brains of experimental animals [72]. AßP is a hydrophobic peptide with an intrinsic tendency to self-assemble and form SDS-stable oligomers in aqueous solution. The monomeric form of AßP has a random coiled structure. Oligomeric AßPs have ß-pleated sheet structures and form insoluble aggregates, termed amyloid fibrils. The neurotoxicity of AßP (1–40) peptides is enhanced by the process of 'aging' (aggregated under incubation at 37°C for several days), as compared to freshly prepared AßP (1–40), and correlates with its ß-sheet contents [73, 74]. Jarrett and Lansbury demonstrated that AßP formed oligomers *via* a nucleation-dependent process and that AßP (1-42) forms seeds for the aggregation of AßP (1-40) [75]. These results suggest that the ratio of AßP (1-40) and AßP (1-42) is crucial in the pathogenesis. The point mutations of APP are located in near the γ-secretase cleavage-site of AßP (1-42) and influences the ratio of AßP (1-40) and AßP (1-42). Other mutations in presenilin genes influence the ratio of AßP (1-40) and AßP (1-42) by causing an increased production of AßP (1–42) in the transfected cell lines [76].

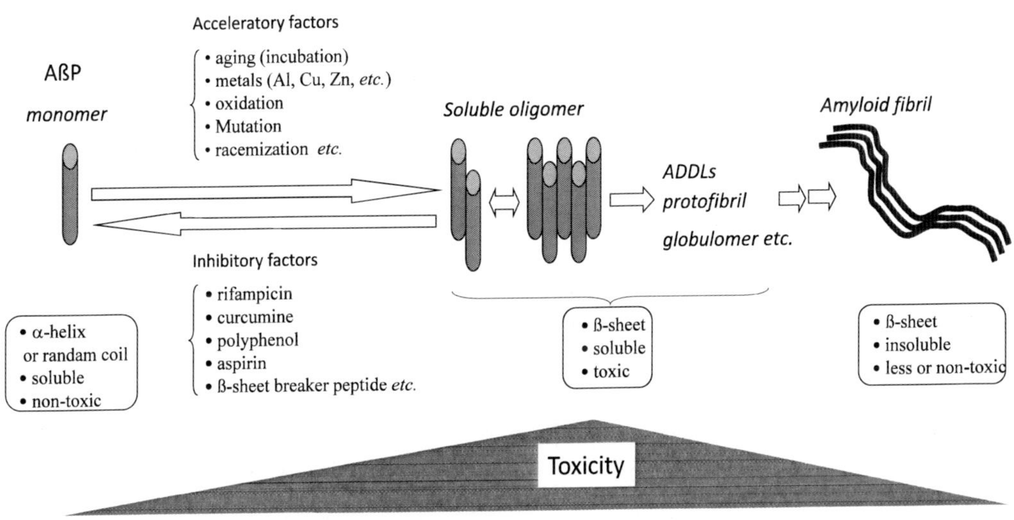

Figure 4. Oligomerization of AßP and its neurotoxicity.
AßP monomers form random-coil structures. However, under aging conditions or in the presence of trace metals such as Al, Zn, and Cu, AßP self-aggregates and oligomerizes (dimmer to protofibrils), and then forms insoluble amyloid fibrils. Although monomeric AßPs are non-toxic, oligomeric AßPs markedly induce neuronal death.

Aluminum and Amyloid Hypothesis

Considering that AßP is secreted in the cerebrospinal fluid (CSF) of young individuals as well as in that of aged individuals and in dementia patients [77], factors that accelerate or inhibit oligomerization may play essential roles in the pathogenesis of AD.

Several factors, such as the concentration of peptides, pH or composition of solvents, and temperature, can influence the oligomerization processes. Oxidations, mutations, and racemization of AßP enhance its oligomerization. Substances that can influence the oligomerization process, such as cholesterol or its oxidation products, apolipoprotein E, transthyletin, rifampicin, curcumin, and aspirin, and ß-sheet breaker peptide have been reported to inhibit AßP oligomerization *in vitro* [78].

Among the acceleratory factors, Al and other trace elements are particularly crucial. Interestingly, rodent AßP exhibits a reduced tendency for oligomerization as compared to human AßP *in vitro,* and the accumulation of AßP is rarely observed in the brains of rodents (rats or mice) as compared to primates (humans or monkeys) [79]. As shown in Figure 3, the amino acid sequences of human and rodent AßP are similar, and rodent AßP differs at only 3 amino acids (Arg^{5}, Tyr^{10}, and His^{13}) from primate AßP. Considering that these three amino acids (Arg^{5}, Tyr^{10}, His^{13}) are able to bind metals, trace metals such as Al, Zn, Cu and Fe may play important roles in the accumulation of AßP in the human brain.

Exley et al. first demonstrated, using CD spectroscopy, that Al induces a conformational change in AßP (1-40) [80]. Al has also been shown to promote the aggregation of ^{125}I-labelled AßP (1-40), with similar findings as for Fe and Zn [81]. We have developed a system for investigating AßP polymerization that involves immunoblotting and precipitation [82]. Using this system, we have shown that Al enhances the polymerization of AßP (1-40) and forms SDS-stable oligomers *in vitro*. The aggregated AßP (1-40) is re-dissolved by adding an Al chelator (DFO). The polymerization induced by Al is more marked than that induced by other metals, including Zn, Fe, Cu, and Cd. Furthermore, we have demonstrated that Al-aggregated AßPs bind tightly to the surface of cultured neurons and form fibrillar deposits, meanwhile, Zn-aggregated AßPs are rarely observed on the surface of cultured neurons [83]. Indeed, Al has been shown to inhibit the degradation of AßP due to conformational changes, and to enhance the accumulation of AßP, as described previously. Other trace metals such as Zn and Cu, enhance AßP oligomerization. Bush et al. have demonstrated that Zn^{2+} at a concentration similar to that in the CSF causes the aggregation of AßP [84]. They also reported that Cu^{2+} binds to AßP, induces its aggregation, and causes a rise in reactive oxygen species levels [85]. Furthermore, APP possesses Cu/Zn binding sites in its amino-terminal domain as well as in the AßP domain. Thus, chelation therapy with clioquinol or other chelators has been proposed as a potential therapy for AD [86].

Recent approaches using size-exclusion chromatography, gel electrophoresis, and atomic force microscopy have demonstrated that not all oligomers are equally neurotoxic. Several stable types of oligomers have been reported: naturally occurring soluble oligomers (dimmers or trimers), AßP-derived diffusible ligands (ADDLs), AßP globulomers, and protofibrils [87]. Hartley et al. separated aggregated AßP (1–40) into low–molecular-weight (mainly monomers), protofibrillar, and fibril fractions using size-exclusion chromatography, and found that the protofibrillar fraction caused marked changes in the electrical activity of cultured neurons and was neurotoxic, but fibrils did not [88]. Walsh et al. found SDS-stable oligomers exist in the conditioned medium with cultured cells exogeneously expressing the human APP gene [89]. The natural AßP oligomers obtained from the CSF of AD patients induce the loss of dendritic spines and synapses, and the blockage of LTP [90].

Morphological analysis using atomic force microscopy demonstrated that the characteristics of AßP oligomers formed in the presence of Al, Zn Cu, and Fe are identical [91]. Sharma et al. revealed that Zn-aggregated AßPs are less toxic than Cu-aggregated AßP [92]. Meanwhile, Al-aggregated AßPs reportedly caused cytoskeletal changes, mitochondorial dysfunction, and increased the production of reactive oxygen species [93]. Bolognin et al. investigated the aggregation and toxicity of AßP induced by Al, Cu, and Fe and Zn [94]. They demonstrated that Al-aggregated AßPs induced an overproduction of APP and tau, but AßP oligomers formed in the presence of other metals did not.

INTERACTIONS OF ALUMINUM WITH OTHER METALS

Influence of Aluminum on Iron Homeostasis

One metal ion usually shares the binding sites of other metal ions, although their binding constants differ. Al binds to various metal-binding proteins and influences metal homeostasis. Thus, the interactions between Al and other metals are important, since the role of these proteins play in metal homeostasis should be considered.

Al has characteristics similar to iron (Fe) and binds to Fe-binding proteins such as ferritin, transferrin, and iron-regulatory protein (IRP) or to Fe chelators such as desferoxamine (DFO). The iron responsive element/iron regulatory protein (IRE/IRP) network regulates the production of Fe binding proteins and prevents the formation of free Fe^{2+} and toxic free radicals [95]. In Fe-deficient conditions, IRP binds to the IRE present in mRNAs such as those encoding ferritin or transferrin, regulating their expression. As the concentration of free Fe^{2+} increases, the binding of Fe to IRP and expression of transferrin down-regulated and that of ferritin is upregulated, thereby decreasing the amount of free Fe^{2+}. Al^{3+} also binds to IRP [96, 97], and thus possibly influences theexpression of Fe-binding proteins encoded by genes containing IREs, thereby causing elevation in Fe concentration. Furthermore, Al influences the uptake of Fe into cultured neurons or glial cells [98.99]. Thus, Al affects Fe homeostasis and the expression of various Fe-regulated genes containing IREs. Al^{3+} also stimulates Fe-induced membrane lipid peroxidation and causes oxidative damage *in vitro* and *in vivo,* although Al^{3+} does not directly affect peroxidation [100, 101]. Everett et al. reported that AßP has the activity to reduce Fe^{3+} to redox-active Fe^{2+}, and found that Al enhanced the reductive activity of AßP [102]. Rogers et al. revealed that APP mRNA also contains an IRE domain as well as ferritin, and its expression is regulated by Fe [103]. Indeed, numerous studies demonstrated that Al caused elevated expression of APP in experimental animals [104-108]. There are other important findings implicating Fe homeostasis in AD pathogenesis. Interestingly, Fe-related genes, such as transferrin C2 or the hemochromatosis gene, were revealed to be risk factors for AD [109]. Imagawa et al. reported that Fe supplementation is effective for the recovery of cognitive functions in AD patients [110].

Influence of Aluminum on Zinc Homeostasis

It is not plausible that Al^{3+} directly interacts with zinc (Zn^{2+}). However, increasing evidence suggests that Al influences Zn homeostasis. Zn is abundantly present in the brain, particularity in the hippocampus and cerebral cortex. A considerable amount of Zn is stored in synaptic vesicles, and is secreted into the synaptic clefts upon excitation of neurons [111]. Zn is an essential element for life and is crucial for memory formation [112]. However, excess Zn is neurotoxic, and has been implicated in the pathogenesis of transient global ischemia and vascular-type senile dementia [113, 114]. Furthermore, Zn is implicated in the pathogenesis of AD [115]. Several studies reported that Zn attenuates the neurotoxicity of Al^{3+} [116, 117] In contrast, we have demonstrated that Al^{3+} attenuates Zn-induced apoptotic death of cultured neurons [118].

Influence of Aluminum on Calcium Homeostasis

Calcium (Ca^{2+}) is essential for the maintenance of normal brain function. Our own and numerous other studies have suggested that dyshomeostasis of Ca^{2+} occurs in AD and other neurodegenerative diseases [119-121]. Al^{3+} blocks various Ca^{2+} channels and influences Ca^{2+} homeostasis [122]. Increasing evidence suggests that Ca^{2+} dyshomeostasis is related with Al-induced neurotoxicity. Indeed, Al neurotoxicity was attenuated by Ca^{2+} channel blockers [123, 124]. We found that Al caused apoptotic death of cultured hippocampal neurons, while brain-derived neurotrophic factor (BDNF) attenuated Al-induced neuronal death [21]. Al inhibits the increase in Ca^{2+} levels that is induced by BDNF; however, it did not influence the glutamate-induced increase in Ca^{2+} levels [125].

ASSOCIATION BETWEEN OTHER DISEASE-RELATED PROTEINS WITH METAL IONS

Recent neurochemical evidence indicated that the oligomerization of proteins and the formation of ß-sheet structures are not restricted to AD, but are also seen in other neurodegenerative diseases such as prion diseases and dementia with Lewy bodies (DLB). The structure of the disease-related proteins, such as AßP in AD, prion protein in prion diseases, α-synuclein in DLB, are identical (Table 2). However, all of these proteins (termed amyloidogenic proteins) share common characteristics in terms of the formation of amyloid fibrils with ß-sheet structures, and in the exhibition of cytotoxicity. These similarities led to the proposal of a new concept, termed "conformational disease", which suggests that a protein's conformation is an important determinant of its toxicity, and consequently, the development of the related disease [126]. Our and numerous other studies have demonstrated that all of these disease-related proteins are directly incorporated into the surface of the cellular membrane and create unregulated cytotoxic pore-like channels that have cytotoxic effects [127-129]. These channels, termed amyloid channels, are giant multilevel pores that can facilitate the transport of large amounts of Ca^{2+}. Their activity is blocked by Zn^{2+} ions [130, 131]. We found that AβP, as well as a PrP fragment (PrP106-126) and an α-synuclein fragment (non-amyloid component; NAC) induced abnormal increases in $[Ca^{2+}]_i$ using a high-resolution multi-site video imaging system along with a Ca^{2+}-sensitive fluorescent dye (fura-2) [132]. These similarities between the amyloidogenic proteins suggest the existence of the common mechanism underlying these neurodegenerative diseases. Furthermore, recent studies have indicated that these amyloidogenic proteins are metalloproteins which possess metal-binding domains, and are implicated in the maintenance of metal homeostasis.

APP has distinct binding domains for Cu, Zn, and Fe, although its physiological roles still remain elusive. However, knock out of APP or its analogue APLP2 in mice resulted in changes in metal distribution [133]. Wong et al. found that APP does not possess ferroxidase but binds to ferroportin, an Fe transporter [134]. Therefore, APP regulates Fe homeostasis, and *vice versa*, Fe controls APP expression as described previously. When this homeostasis is disrupted, increased Fe or APP will trigger the degenerative processes. A decreased level of ferroportin and an accumulation of Fe in the AD brain have been reported [135]. Additionally, another trace metal, Cu also influences the expression of APP [136].

Table 2. Characteristics of amyloidogenic proteins and the related peptides

Disease	The primary sequence of amyloidogenic protein or its fragment peptide	Metal	ß-sheet formation	Cyto-toxicity	Channel formation
Alzheimer's disease	*AßP (1-42)* *DAEFRHDSGYEVHHQKLVFFAEDVGSNKGAIIGLMVGG VVIA*	Al, Zn Cu, Fe	+	+	+
Prion disease	Prion protein: *PrP106-126* MANLGCWMLVLFVATWSDLGLCKKRPKPGGWNTG GSRYPGQGSPGGNRYPPQGGG GWGQPHGGGWGQPHGGGWGQPHGGGWGQPHGGG WGQGGGTHSQWNKPSKP*KTNMK HMAGAAAAGAVVGGLG*GYVL GSAMSRPIIHFGSDYEDRYYRENMHRYPNQVYYRPM DEYSNQNNFVHDCVNITIKQHTVTTTTKGENFTETD VKMMERVVEQMCITQYERESQAYYQRGS SMVLFSSPPVILLISFLIFL IVG	Zn, Cu Mn Fe	+	+	+
Dementia with Lewy bodies (DLB)	-synuclein; *NAC* (a *fragment of □-synuclein)* MDVFMKGLSKAKEGVVAAAEKTKQGVAEAA GKTKEGVLYVGSKTKEGVVHGVTTVAEKTK*EQVSN VGGAVVTGVTAVAHKTVEGAGNFAAATGLV*KKDQKNE SGFGPEGTMENSENMPVNPNNETYEMPPEEEYQDYD PEA	Cu Fe Al Mn	+	+	+
Triplet- repeat disease	*Polyglutamine* MATLEKLMKAFESLKSF*QQQQQQQQQQQQQQQQQ QQQQQQ*PPPPPPPPPPPQLPQPPPQAQPLLPQPQPPPPP PPPPPGP	Fe	+	+	+
Diabetes mellitus	*Human amylin* KCNTATCATQRLANFLVHSSNNFGAILSSTNVG SNTY	Cu Al	+	+	+

* The sequence of fragment peptide of each amyloidogenic protein (PrP106-126, NAC, polyglutamine) is indicated in italic form.

Prion diseases, including human kuru, Creutzfeldt-Jakob disease, and bovine spongiform encephalopathy (BSE), are associated with the conversion of a normal prion protein (PrP^{C}) into an abnormal scrapie isoform (PrP^{SC}) [137]. The ß-sheet region of PrP^{SC} has been suggested to play a crucial role in its transmissible degenerative processes. Although the physiological roles of PrP^{C} is still controvercial, increasing evidence suggests that PrP^{C} is a metalloprotein and implicates in the metal homeostasis [138]. PrP^{C} contains 208 amino acid residues and possesses a highly conserved octarepeat domain composed of multiple tandem copies of the 8-residue sequence PHGGGWGQ in its N-terminal. In 1997, Brown et al. reported decreased levels of Cu in the brains of PrP-knockout mice as compared with those of normal mice [139]. Jackson et al. reported that PrP^{C} binds four Cu atoms at its octarepeat domain and binds two additional Cu atoms to two other histidine (His) residues, His^{96} and His^{111} [140]. They also demonstrated that other metal ions including Zn^{2+}, Mn^{2+}, and Ni^{2+} can bind to these sites, although with lower affinity than Cu^{2+}. PrP^{C} reportedly transports Cu from the extracellular space to the intracellular space *via* endocytosis and regulates the intracellular concentration of Cu [141]. PrP^{C} possesses or modulates Cu/Zn superoxide dismutase (Cu/ZnSOD) activity in the brain, and has protective roles against oxidative stress. Recent studies have suggested that PrP^{C} regulates the function of the *N*-methyl-D-aspartate (NMDA)-type glutamate receptor in a Cu-dependent manner.

PrP^{C} has the next higher affinity for Zn^{2+}, after Cu^{2+}. Recent bioinformatics analysis has revealed evolutionary similarities between prion genes and genes encoding the Zrt-, Irt-like protein (ZIP) Zn transporters [142]. There are two major Zn transporter families (the ZnT family and the ZIP family) that control cellular Zn homeostasis in mammals [143]. ZIP Zn transporters primarily serve to move Zn into the cytoplasm. ZIP proteins are encoded by the *SLC39a* gene family, which consists of 14 genes in humans and mice. Sequence similarities have been reported between the genes encoding PrP^{C} and ZIP5, ZIP6, and ZIP 10. PrP^{C} reportedly co-localizes with ZIP 5 in neuroblastoma cells. Watt et al. found that PrP^{C} facilitates the uptake of Zn into neuronal cells, which is mediated by α-amino-3-hydroxy-5-methyl-4-isoxazolepropionate (AMPA) receptors, and hypothesized that PrP^{C} acts as a Zn sensor, detecting extracellular levels up to a certain threshold [144]. PrP^{C} attenuates Zn-induced neurotoxicity. In addition to Cu and Zn, other metals, such as Fe and Mn, are associated with PrP. There have been several studies suggesting that Mn may facilitate prion diseases. The levels of trace elements in prion-infected hamster brains using X-ray photoelectron emission microscopy with synchrotron radiation revealed the reduced Cu and increased Mn in plaques composed of prion protein. Singh et al. have suggested that PrP^{C} acts as a ferrireductase that is responsible for reducing Fe^{3+} to bioavailable Fe^{2+} and modulates the cellular uptake of Fe [145]. PrP is co-transported with ferritin, an Fe binding protein. Decreased levels of transferrin, an Fe transporter protein, have been observed in the CSF of CJD patients.

The oligomerization and fibrillation of α-synuclein has been implicated in the formation of abnormal cellular inclusions, termed Lewy bodies, and the etiology of DLB [146]. Non-amyloid component (NAC), a fragment peptide of α-synuclein, accumulates in AD senile plaques and causes apoptotic neuronal death. Uversky et al. found the metals, such as Al and Mn, enhance the oligomerization of α-synuclein [147]. α-Synuclein reportedly binds Fe and acts as a ferrireductase, that reduces Fe^{3+} to bioavailable Fe^{2+} [148]. The mRNA of α–synuclein also has an IRE domain, and Fe regulates its expression [149].

Hypothesis: Crosstalk of Aluminum, Other Metals, and Amyloidogenic Proteins in the Synapse

AßP is a synaptotoxin and the loss of synapses forms part of the neurotoxicity process. Neurodegeneration in prion disease and in DLB has been observed to involve in the synapses. Thus, the synapse is considered to be the major target of neurotoxicity in these diseases. It is widely recognized that APP is present in the pre-synaptic domain, and AßP is secreted into synaptic clefts. PrP^C is localized in synapses and colocalizes with receptors, such as the NMDA-type glutamate receptor or the AMPA-type glutamate receptor, in postsynaptic membranes. α-synuclein is present in the pre-synaptic domain. Considering the small volume of the synaptic cleft (20-nm height and 120-nm radius) [150], it is highly likely that these amyloidogenic proteins interact with other proteins or receptors, including each other, in this region. Indeed, PrP^C reportedly regulates AßP levels by influencing the cleavage of APP, and attenuates the neurotoxicity of AßP oligomers [151, 152]. Moreover, α-synuclein fragment co accumulates with AßP in senile plaques [153].

Furthermore, a large amount of Zn^{2+} is also present in synaptic vesicles, along with glutamate, and is secreted into synaptic clefts upon neuronal excitation. Although the level of Zn in the synaptic cleft is still controversial, some findings demonstrated concentrations of up to 300 μM of Zn exist in this region. Therefore, it is assumed that the synaptic cleft provides a basis for functioning of these amyloidogenic proteins in a Zn pool.

Considering the results of our own together with those of other studies, we propose the following hypothetical scheme for the involvement of trace elements and amyloidogenic proteins in the synapse. Under normal condition (Figure 5A), these amyloidogenic proteins colocalize in the synapse and have cytoprotective effects, influencing the influx of metals and controlling their homeostasis, and serve to maintain normal brain functions. APP modulates Cu or Zn, and regulates Fe homeostasis in the synapse. Both of PrP^C and α-synuclein have ferrireductase activity, catalyzing Fe^{3+} to Fe^{2+}. These proteins therefore regulate the Fe levels, and particularly Fe^{2+}/Fe^{3+} ratio. Fe^{2+} is necessary for the synthesis of neurotransmitters, such as dopamine, but excessive Fe^{2+} causes oxidative stress and neurotoxicity; thus, the ratio of Fe^{2+} and Fe^{3+} is important for normal brain function. Moreover, the disruption of Fe homeostasis triggers the over expression of APP or α-synuclein and the accumulation of AßP. Zn is crucial for memory formation, in addition to Ca. However, excessive Zn levels are toxic and are supposedly central to the ischemia-induced neuronal death and the pathogenesis of vascular dementia. Therefore, Zn homeostasis in synaptic clefts is essential for proper brain function. To regulate the concentration of Zn in synapses, Zn transporter 1 (ZnT-1) is localized to the post-synaptic domain, where it exports Zn [154]. PrP^C acts as a Zn sensor and regulates its concentration in synaptic clefts as mentioned before, and is critical for the regulation of Zn homeostasis. Furthermore, PrP^C also regulates Cu levels in the synapse, and thereby controls APP expression given that Cu influences APP expression. PrP^C also has SOD activity, and protects cells from free radicals.

In pathological conditions (Figure 5B), Al^{3+} entering into the brain from the external environment can cause various adverse effects as noted earlier. Al^{3+} has similar characteristics with Fe^{3+}, and causes Fe dyshomeostasis. For example, Al^{3+} binds to the APP mRNA and enhances its expression, which eventually results in the upregulation of AßP. Al^{3+} also causes the aggregation of AßP, enhances its accumulation, and finally enhances its toxicity. When Al perturbs the ratio of Fe^{2+} and Fe^{3+}, it enhances Fe^{2+}-induced free radical formation, and thereby neurotoxicity.

Al^{3+} also impairs Zn homeostasis and Ca homeostasis. The dishomeostasis of these metals in the synaptic region causes synaptic degeneration, and thusneuronal degeneration, which is central to the pathogenesis of these neurodegenerative diseases. The substances that regulates Zn homeostasis, such as carnosine and MT-3, are therefore important [155, 156].

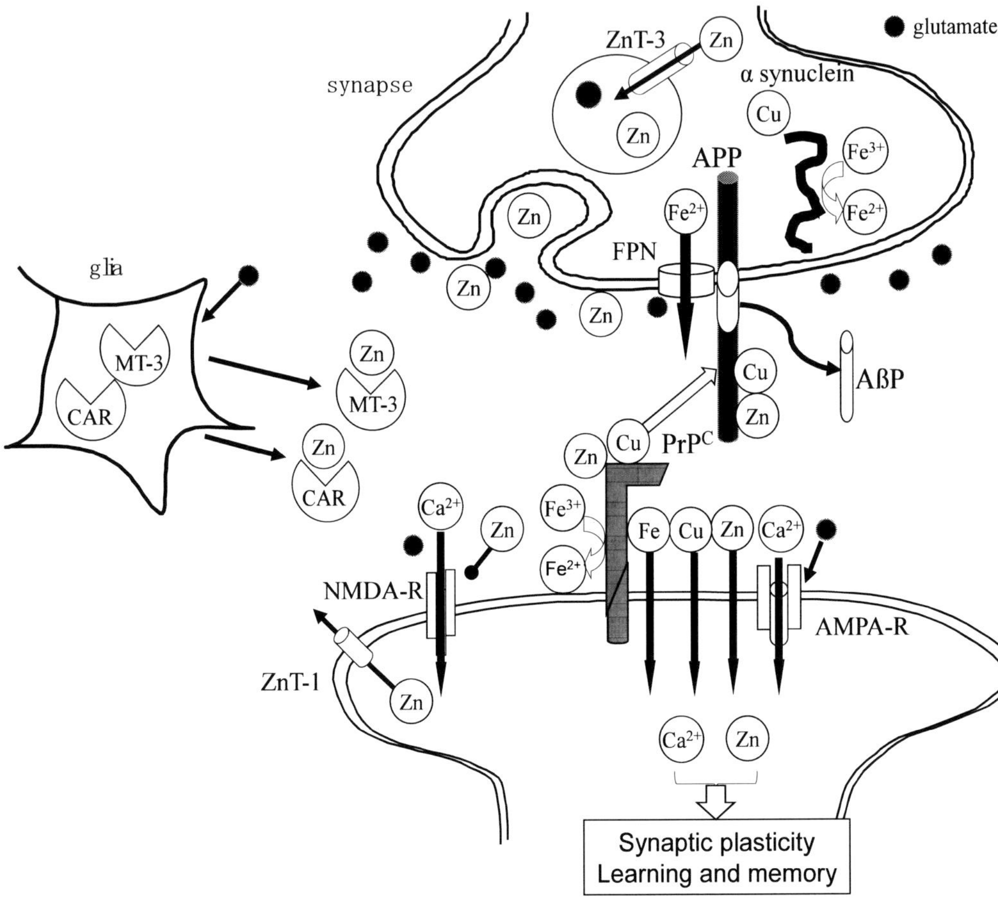

Figure 5A. Physiological roles of amyloidogenic proteins in the synapse.
APP and α-synuclein exists in the pre-synaptic domain, and PrP^C localizes in the post-synaptic domain of synapses. Zn and glutamate are accumulated in synaptic vesicles and are released into synaptic clefts during neuronal excitation. Zn^{2+} regulates Ca^{2+} influx through glutamate receptors (NMDA-R, Ca-A/K), modulates neuronal information, and is implicated in the maintenance of synaptic plasticity and memory formation, as is Ca^{2+}. The level of Zn^{2+} is controlled by ZnT-1, metallothionein-3, and carnosine. PrP^C binds to Ca-A/K and regulates Zn^{2+} level. Moreover, PrP^C has SOD activity and also regulates Cu^{2+} level, which influence the sequencing of APP. PrP^C and α-synuclein have ferrireductase activity, which convertts Fe^{3+} to Fe^{2+}. APP binds to feroportin and thereby regulates Fe^{2+} influx. Therefore, these proteins play cytoprotective roles including the regulation of trace element homeostasis and the protection from free radicals in the synapse.
ZnT-1: zinc transporter 1, AMPA-R: AMPA-type glutamate receptor, NMDA-R: NMDA-type glutamate receptor; FPN: ferroportin.

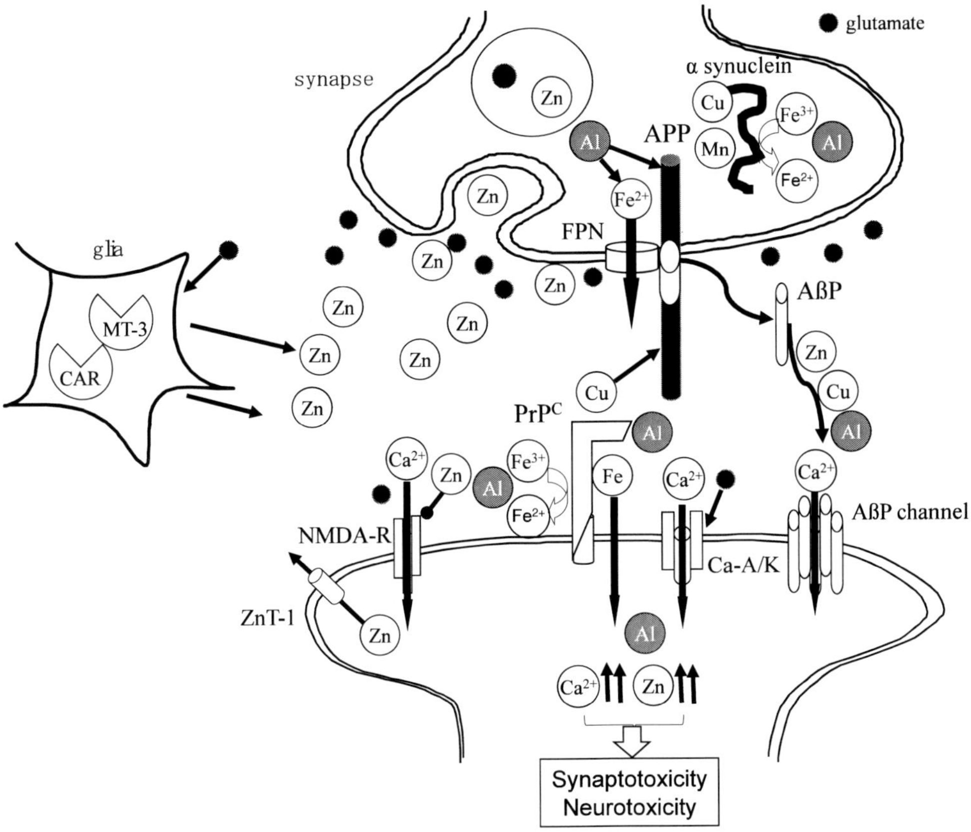

Figure 5B. Al-induced dyshomeostasis of metals in the pathogenetic condition
In the pathological conditions, when Al accumulates in the brain, Al can influence numerous pathways in the synapse. Al enhances APP expression, influences Fe-related pathways, disrupts Fe homeostasis, and enhances the formation of free radicals. Dyshomeostasis of Fe triggers the upregulation of APP and the over-production of AßP. Al causes the oligomerization and accumulation of AßP, and enhances its neurotoxicity. Al also influences Ca^{2+} homeostasis as a Ca^{2+} channel blocker. Finally, the dyshomeostasis of these proteins and trace elements leads to loss of the cytoprotective functions of these proteins, degeneration of synapses, degeneration of neurons, and finally the pathogenesis of neurodegenerative diseases.

This working hypothesis may be useful in developing an understanding of the link between AD and trace elements, including Al, Zn, Cu, and Fe. Considering the implications of metals in AD pathogenesis, chelation therapy for AD treatment is of great interest.

Bioavailability of Aluminum: From Mouth to Brain

Even if the link between Al and AD is controversial, as noted previously, it is clear that Al is toxic to the CNS and causes dementia when present in the brain. Therefore, the level of Al in our body, namely its intake and bioavailability, is crucial to our health [157]. The primary source of Al taken in by humans is food stuffs. Typically, the level of Al in food is low and varies within a wide range. However, several plants have been reported to retain Al; thus, the food stuffs derived from such plants would contain high Al levels. Furthermore, the

Al in food additives or cooking utensils can result in contamination of foods and can enhance the intake of Al. The typical daily intake of Al by humans is estimated to be 10-20 mg/day. In 1989, a joint FAO/WHO Expert Committee on Food Additives (JECFA) recommended a provisional tolerable weekly intake (PTWI) of 7.0 mg/kg body weight Al; however, this was changed to 1.0 mg/kg body weight in 2007 because of its potential effects on the reproductive system and the developing nervous system. This value was further changed to 2.0mg/kg body weight in 2011 since the difficulty in the limitation of Al intake.

In general, the intake of Al does not correlate with the amount of Al in the body. Contributing to this is the absorption of metals, including Al, by the gastrointestinal tract, which is low and widely variable [158]. Thus, the bioavailability of Al, corresponding to the amounts of Al absorbed by the gastrointestinal tract, is crucial. Research into the bioavailability of Al has previously been restricted by a lack of suitable radioactive Al isotopes. Recent developments in acceleration mass spectrometry using ^{26}Al have advanced such research. Usually, only a small amount (approximately less than 1%) of the Al present in food is absorbed *via* the gastrointestinal pathway. However, the uptake of Al through this pathway is complex and is influenced by various factors including an individual differences, age, pH, stomach contents, chemical speciation of Al, and coexistence substances. For example, Al hydroxide ($Al(OH)_3$), a key component of antacids, is absorbed to a lesser extent than Al citrate. This result may explain why the usage of antacids is not associated with AD pathogenesis. The coexistence of organic ligands can promote the absorption of Al, as is the case with citric acid and maltol. Moreover, silicates prevent the absorption of Al. Further illustrating the lack of correlation between Al intake and body levels of Al is the fact that, while Al present in tea can be a major contributor to daily intake, Al in tea infusion is less well absorbed. Conversely, it seems that Al eluted from cooking utensils is highly bioavailable [159]. Discontinuation of the use of Al utensils has been shown to induce a decrease in serum Al levels. The absorption rates of Al reportedly increase in aged people, in patients with Down's syndrome, and in patients with AD.

Once absorbed from the gastrointestinal tract, Al appears rapidly in the blood, with approximately 80% bound to transferrin. The remaining 20% of Al in the blood is bound to albumin and citrate. Around 50% of the Al absorbed into the serum is excreted in the urine, with most of what remains accumulating in bone. A small, but considerable amount of Al crosses the blood-brain barrier, possibly *via* the transferrin-receptor pathway or monocarboxylate transporters. Once entered into the brain, Al is retained, accumulating in a semi-permanent manner [160]. This was demonstrated by Kobayashi et al., who found that intraperitoneally or orally administered ^{26}Al was transferred into the brain, where the amount of Al remained unchanged for 35 days, despite a rapid disappearance of the metal from the blood [161]. It thus appears that any elimination of Al from the brain occurs at an extremely slow rate. These results are consistent with an age-dependent increase in brain Al content, a relationship not observed for other trace elements.

Particular consideration should be given to the use of Al in medicine, in which context the bioavailability is typically high. As noted, compared with the iatrogenic absorption of Al from dialysis solution, the gastrointestinal absorption of Al is low. A considerable amount of Al has been shown to be contaminated in total parenteral nutrition (TPN) [162]. The Al in TPN solutions is completely absorbed into the blood flow and while a portion of this is excreted via the kidneys, some patients receiving TPN solutions experience kidney failure. This can also be an issue for infants, in whom the kidneys are not yet fully functional. Accumulation of Al in the bones of patients receiving TPN has been reported. Furthermore, preterm infants receiving TPN containing a high concentration of Al had lower scores for mental development than age-matched infants receiving TPN containing low Al levels. These findings have resulted in the U.S. Food and Drug Administration (FDA), the North American Society for Pediatric Gastroenterology and Nutrition, and other related societies recommending a reduction in the contamination of Al in TPN solutions. The FDA has published a final rule requiring that the concentration of Al in TPN should be indicated on the label at 2003 [163]. In Japan, a convenient method using HPLC for determination of Al in TPN was added in the 16th Japanese Pharamacopoeia, since there was a need for the development of methods for analysis of Al analysis in TPN solutions.

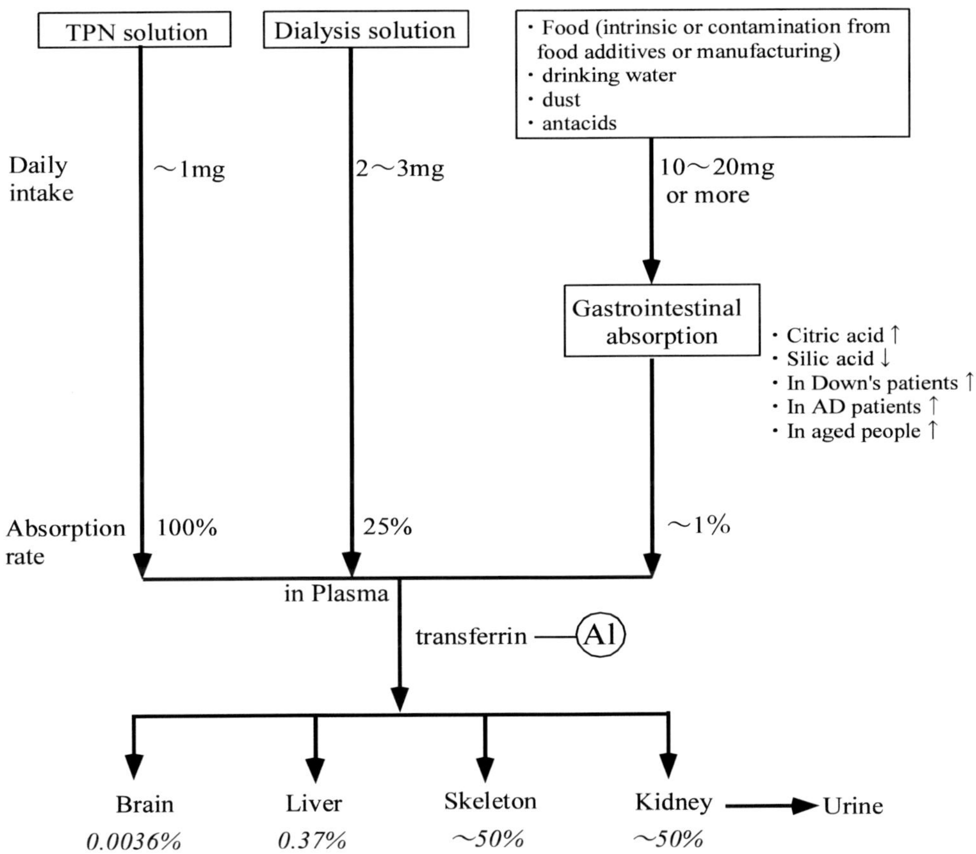

Figure 6. Biological fate of aluminum.
Daily intake of Al from foods is estimated to be 10-20mg derived from intrinsic foodstuffs, contamination by food additives or utensils, etc. Although the intake from antacids will contribute much more, its absorption rate is usually low in healthy adults. However, antacid could cause adverse effects in dialysis patients or patients with kidney failures. In general, gastrointestinal absorption rate is approximately 1%. However, it is different in each individual, and largely influenced by age, pH, stomach contents, chemical speciation of Al, coexistence substances such as silic acid. Meanwhile, Al in the dialysis solution or in TPN solution could be easily absorbed and will contribute more to total amount of Al in the body. Once entered in the blood flow, approximately 50% of Al in the plasma is excreted from kidney. However, rest is accumulated in the skeletons permanently. Furthermore, small but considerable amount of Al passes through blood brain barrier and enters into the brain. Al in the brain is almost not excreted and accumulated through lifetime. Values of the absorption rate are obtained from Jouhanneau et al. (1997).

There is a comparatively very high amount of Al in antacids; the prolonged use of these agents may thus increase the risk of encephalopathy in patients with renal failure or kidney diseases. Indeed, the Japanese Ministry of Health, Labor and Welfare has recommended that patients on dialysis or with kidney failure should not use commercial Al-containing antacids in 2002.

Other sources of Al for infants are milk and formula, in which Al is a known contaminant [164]. Yumoto et al. demonstrated a transfer of ^{26}Al administered to mother rats into the brains of suckling infants *via* maternal milk [165]. The absorbed Al remained unchanged through the lifetime of the infant rats. These results indicate that Al is readily transferred across the blood-placental barrier, whereby it can subsequently enter the fetal or infant brain. As the *in utero* period is critical to brain development, it is most likely that Al entering the brain during this time will have adverse effects.

CONCLUSION

In this review, we have summarized the properties associated with various aspects of Al neurotoxicity. There is growing evidence for a link between Al and AD, and between other metals and AD. Nevertheless, because the precise mechanism of AD pathogenesis remains unknown, this issue is still controversial. However, it is widely accepted that Al is a recognized neurotoxin, and that it could cause cognitive deficiency and dementia when it enters the brain and may have various adverse effects on the CNS. In general, the absorption of metals by the gastrointestinal tract is widely variable and is influenced by various factors including an individual differences, age, pH, and stomach contents. Recent studies using mass spectrometry of ^{26}Al have demonstrated that a small, but a considerable amount of Al crosses the blood brain barrier, enters into the brain, and accumulates in a semi-permanent manner. Therefore, Al can cause severe health problems in particular populations, including infants, elderly people, and patients with impaired renal functions, and unnecessary exposure to Al should be avoided for such patients. The characteristics of Al neurotoxicity are complex, and further research is needed, particularly in relation to bioavailability, cellular effects, metabolism, and metal-metal interactions.

REFERENCES

[1] Spofforth, J, Edin, l.R.C. and Eng, M.R.C. 1921. Case of aluminium poisoning. *Lancet,* 1: 1301.

[2] Chusid, J.G., Pacella, B.L., Kopeloff, L.M. and Kopeloff, N. 1951. Chronic epilepsy in the monkey following multiple intracerebral injection of alumina cream. *Proc. Soc. Exp. Biol. Med.* 78: 53-4.

[3] Wills, M.R. and Savory, J. 1958. Aluminum and chronic renal failure: sources, absorption, transport, and toxicity. *Crit. Rev. Clin. Lab. Sci.* 27: 59-107.

[4] Alfrey, A.C., LeGendre, G.R. and Kaehny, W.D. 1976. The dialysis encephalopathy syndrome: Possible aluminium intoxication. *N. Engl. J. Med.* 294: 184-8.

[5] Altmann, P., Cunningham, J., Dhanesha, U., Ballard, M., Thompson, J., Marsh, F. 1999. Disturbance of cerebral function in people exposed to drinking water contaminated with aluminium sulphate: retrospective study of the Camelford water incident. *Br. Med. J.* 319: 807-11.

[6] Campbell, A. 2002. The potential role of aluminium in Alzheimer's disease, *Nephrol. Dial. Transplant.* 17: 17-20.

[7] McLachlan, D.R. 1995. Aluminium and the risk for Alzheimer's disease. *Environmetrics.* 6: 233-75.

[8] Zatta, P. (eds). 2002. Recent topics in aluminium chemistry. *Coordination Chemistry Reviews.* 228, Issue 2.

[9] Exley, C. 2005. The aluminium-amyloid cascade hypothesis and Alzheimer's disease. *Subcell Biochem.* 38: 225-34.

[10] Kawahara, M. and Kato-Negishi, M. 2011. Link between aluminum and the pathogenesis of Alzheimer's disease: the integration of the aluminum and amyloid cascade hypotheses. *Int. J. Alzheimer Dis.* 276393.

[11] Shiraki, H. and Yase, Y. 1991. Amyotrophic lateral sclerosis and Parkinsonism-dementia in the Kii peninsula: comparison with the same disorders in Guam and with Alzheimer's disease. *Handbook of Clinical Neurology.* 15: 273-300.

[12] Petrik, M.S., Wong, M.C., Tabata, R.C., Garry, R.F. and Shaw, C.A. 2007. Aluminum adjuvant linked to Gulf War illness induces motor neuron death in mice. *Neuromolecular Medicine.* 9: 83-100.

[13] Shaw, C.A., Tomljenovic, L. 2013. Aluminum in the central nervous system (CNS): toxicity in humans and animals, vaccine adjuvants, and autoimmunity. *Immunol. Res.* 56: 304-16.

[14] Lidsky, T.I. 2014. Is the Aluminum Hypothesis dead? *J. Occup. Environ. Med.* 56: S73-9.

[15] Exley, C. 2003. A biogeochemical cycle for aluminium? *J. Inorg. Biochem.* 97: 1-7.

[16] Martin, R.B. 1997. Chemistry of aluminum in the central nervous system. Yasui M, et al. (eds): *Mineral and Metal Neurotoxicology*. CRC Press, N.Y., pp75-81.

[17] Lukiw, W.J., LeBlanc, H.J., Carver, L.A., McLachlan, D.R. and Bazan, N.G. 1998. Run-on gene transcription in human neocortical nuclei. Inhibition by nanomolar aluminum and implications for neurodegenerative disease. *J. Mol. Neurosci.* 11: 67-78.

[18] Lukiw, W.J. 2012. NF-κB-regulated, proinflammatory miRNAs in Alzheimer's disease. *Alzheimers Res Ther.* 4:47. doi: 10.1186/alzrt150.

[19] Julka, D. and Gill, K.D. 1996. Involvement of altered cytoskeletal protein phosphorylation in aluminum-induced CNS dysfunction. *J. Biochem. Toxicol.* 11: 227-33.

[20] Diaz Nido, J. and Avila, J. 1990. Aluminum induces the in vitro aggregation of bovine brain cytoskeletal protein. *Neurosci. Lett.* 110: 221-6.

[21] Kawahara, M., Kato-Negishi, M., Hosoda, R., Imamura, L., Tsuda, M. and Kuroda, Y. 2003. Brain-derived neurotrophic factor protects cultured rat hippocampal neurons from aluminum maltolate neurotoxicity. *J. Inorg. Biochem.* 97: 124-31.

[22] Guo, G.W. and Liang, Y.X. 2001. Aluminum-induced apoptosis in cultured astrocytes and its effect on calcium homeostasis. *Brain Res.* 888: 221-6.

[23] Platt, B., Carpenter, D.O., Büsselberg, D., Reymann, K.G. and Riedel, G.. 1995. Aluminum impairs hippocampal long-term potentiation in rats in vitro and in vivo. *Exp. Neurol.* 134: 73-86.
[24] Kanazirska, M., Vassilev, P.P., Birzon, S.Y., Vassilev, P.M. 1997. Voltage-dependent effect of Al^{3+} on channel activities in hippocampal neurons. *Biochem. Biophys. Res. Commun.* 232: 84-7.
[25] Yokel, R.A., Allen, D.D. and Meyer, J.J. 1994. Studies of aluminum neurobehavioral toxicity in the intact mammal. *Cell. Mol. Neurobiol.* 14: 791-808.
[26] Kaneko, N., Takada, J., Yasui, H., Sakurai, H. 2007. Memory deficit in mice administered aluminum-maltolate complex. *Biometals.* 19: 83-9.
[27] Selkoe, D.J. 1991. The molecular pathology of Alzheimer's disease. *Neuron.* 6: 487-98.
[28] Terry, R.D., Masliah, E., Salmon, D.P., Butters, N., DeTeresa, R., Hill, R., et al. 1991. Physical basis of cognitive alterations in Alzheimer's disease: synapse loss is the major correlate of cognitive impairment. *Ann. Neurol.* 30: 572-80.
[29] Klatzo, I. Wisniewski, H. and Streicher, E. 1965. Experimental production of neurofibrillary degeneration I. Light microscopic observation. *J. Neuropathol. Exp. Neurol.* 24: 187-99.
[30] Crapper D.R., Krishnan S.S. and Dalton A.J. 1973. Brain aluminium distribution in Alzheimr's disease and experimental neurofibrillary degeneration. *Science.* 180: 511-3.
[31] Martyn, C.N., Barker, D.J., Osmond, C., Harris, E.C., Edwardson, J.A. and Lacey, R.F. 1989. Geographical relation between Alzheimer's disease and aluminum in drinking water. *Lancet.* 1: 59-62.
[32] Kruck T.P. and McLachlan D.R. 1989. Aluminum as a pathogenic factor in senile dementia of the Alzheimer type: ion specific chelation. *Prog. Clin. Biol. Res.* 317: 1155-67.
[33] Wisniewski, H.M. and Wen, G.Y. 1992. Aluminium and Alzheimer's disease. *Ciba Found. Symp.* 169: 142-154.
[34] Doll, R. 1993. Alzheimer's disease and Environmental aluminium. *Age and Ageing.* 22: 138-53.
[35] Harrington, C.R., Wischik, C.M., McArthur, F.K., Taylor, G.A., Edwardson, J.A. and Candy, J.M. 1997. Alzheimer's disease-like changes in tau protein processing: association with aluminium accumulation in brains of renal dialysis patients. *Lancet.* 343: 993-7.
[36] Kawahara, M., Muramoto, K., Kobayashi, K. and Kuroda, Y. 1992. Functional and morphological changes in cultured neurons of rat cerebral cortex induced by long-term application of aluminum. *Biochem. Biophys. Res. Commun.* 189: 1317-22.
[37] Yamamoto, H., Saitoh, Y., Yasugawa, S., Miyamoto, E. 1990. Dephosphorylation of tau factor by protein phosphatase 2A in synaptosomal cytosol fractions, and inhibition by aluminum. *J Neurochem.* 55:683-90.
[38] Scott, C.W., Fieles, A., Sygowski, L.A., Caputo, C.B. 1993. Aggregation of tau protein by aluminum. *Brain Res.* 628: 77-84.
[39] Crowther, R.A. 1991. Straight and paired helical filaments in Alzheimer disease have a common structural unit. *Proc. Natl. Acad. Sci.* 88: 2288-92.
[40] Hardy, J. and Selkoe, D.J. 2002. The amyloid hypothesis of Alzheimer's disease: progress and problems on the road to therapeutics. *Science.* 297: 353-6.

[41] Kuroda, Y., Kobayashi, K., Ichikawa, M., Kawahara, M., Muramoto K. 1995. Application of long-term cultured neurons in aging and neurological research: aluminum neurotoxicity, synaptic degeneration and Alzheimer's disease. *Gerontology.* 41: 2-6.

[42] Campbell, A., Kumar, A., La Rosa, F.G., Prasad, K.N. and Bondy, S.C. 2000. Aluminum increases levels of beta-amyloid and ubiquitin in neuroblastoma but not in glioma cells. *Proc. Soc. Exp. Biol. Med.* 223: 397-402.

[43] Pratico, D., Uryu, K., Sung, S., Tang, S., Trojanowski, J.Q. and Lee, V.M. 2002. Aluminum modulates brain amyloidosis through oxidative stress in APP transgenic mice. *FASEB J.* 16: 1138-40.

[44] Ribes, D., Colomina, M.T., Vicens, P., Domingo, J.L. 2010. Impaired spatial learning and unaltered neurogenesis in a transgenic model of Alzheimer's disease after oral aluminum exposure. *Curr. Alzheimer Res.* 7:401-8.

[45] Exley, C. and Esiri, M.M. 2006. Severe cerebral congophilic angiopathy coincident with increased brain aluminium in a resident of Camelford, Cornwall, UK. *J. Neurol. Neurosurg. Psychiatry.* 77: 877-9.

[46] Dickstein, D.L., Walsh, J., Brautigam, H., Jr Stockton, S.D., Gandy, S., Hof, P.R. 2010. Role of vascular risk factors and vascular dysfunction in Alzheimer's disease. *Mt. Sinai. J Med.* 77: 82-102.

[47] Bjertness, E., Candy, J.M., Torvik, A., Ince, P., McArthur, F., Taylor, G.A., et al. 1996. Content of brain aluminum is not elevated in Alzheimer disease. *Alzheimer Dis. Assoc. Disord.* 10: 171-4.

[48] Drago, D., Bettella, M., Bolognin, S., Cendron, L., Scancar, J., Milacic, R., et al. 2008. Potential pathogenic role of beta-amyloid(1-42)-aluminum complex in Alzheimer's disease. *Int. J. Biochem. Cell Biol.* 40: 731-46.

[49] González-Domínguez, R., García-Barrera, T., Gómez-Ariza, J.L. 2014. Characterization of metal profiles in serum during the progression of Alzheimer's disease. *Metallomics*. 6: 292-300.

[50] Virk, S.A., Eslick, G.D. 2015. Aluminum levels in brain, serum, and cerebrospinal fluid are higher in Alzheimer's disease cases than in controls: A series of meta-analyses. *J. Alzheimers Dis.* 47: 629-38.

[51] Arain, M.S., Afridi, H.I,, Kazi, T.G., Talpur, F.N., Arain, M.B., Kazi, A. et al. 2015. Correlation of aluminum and manganese concentration in scalp hair samples of patients having neurological disorders. *Environ. Monit. Assess.* 187:10. doi: 10.1007/s10661-014-4172-0.

[52] Landsberg, J.P., McDonald B. and Watt, F. 1992. Absence of aluminium in neuritic plaque cores in Alzheimer's disease. *Nature.* 360: 65-9.

[53] Candy, J.M., McArthur, F.K., Oakley, A.E., Taylor, G.A., Chen, C.P., Mountfort, S.A. et al. 1992. Aluminium accumulation in relation to senile plaque and neurofibrillary tangle formation in the brains of patients with renal failure. *J. Neurol. Sci.* 107: 210-8.

[54] Bouras, C. Giannakopoulos, P. Good, P.F. Hsu, A. Hof, P.R. and Perl, D.P. 1997. A laser microprobe mass analysis of brain aluminum and iron in dementia pugilistica: comparison with Alzheimer's disease. *Eur. Neurol.* 38: 53-8.

[55] Yumoto, S., Kakimi, S., Ohsaki, A., Ishikawa, A. 2009. Demonstration of aluminum in amyloid fibers in the cores of senile plaques in the brains of patients with Alzheimer's disease, *J. Inorg. Biochem.* 103: 1579-84.

[56] Flaten, T.P. 2001. Aluminium as a risk factor in Alzheimer's disease, with emphasis on drinking water. *Brain Res. Bull.* 55: 187-96.

[57] Wettstein, A., Aeppli, J., Gautschi, K., and Peters, M. 1991. Failure to find a relationship between mnestic skills of octogenarians and aluminum in drinking water. *Int. Arch. Occup. Environ. Health.* 63: 97-103.

[58] Martyn, C.N., Coggon, D.N., Inskip, H., Lacey, R.F. and Young, W.F. 1997. Aluminum concentrations in drinking water and risk of Alzheimer's disease. *Epidemiology.* 8: 281-6.

[59] McLachlan, D.R., Bergeron, C., Smith, J.E., Boomer, D. and Rifat, S.L. 1996. Risk for neuropathologically confirmed Alzheimer's disease and residual aluminum in municipal drinking water employing weighted residential histories. *Neurology.* 46: 401-5.

[60] Gauthier, E., Fortier, I., Courchesne, F., Pepin, P., Mortimer, J. and Gauvreau, D. 2000. Aluminum forms in drinking water and risk of Alzheimer's disease. *Environ. Res.,* 84: 234-46.

[61] Birchall J.D. and Chappell, J.S. 1988. The chemistry of aluminum and silicon in relation to Alzheimer's disease. *Clin. Chem.* 34: 265-7.

[62] Edwardson, J.A., Moore, P.B., Ferrier, I.N., Lilley, J.S., Newton, G.W.A., Barker, J.J., et al. 1993. Effect of silicon on gastrointestinal absorption of aluminium. *Lancet.* 342: 211-2.

[63] Jacqmin-Gadda, H., Commenges, D., Letenneur, L., Dartigues, J.F. 1996. Silica and aluminum in drinking water and cognitive impairment in the elderly. *Epidemiology.* 7: 281-5.

[64] 64 Rondeau, V., Jacqmin-Gadda, H., Commenges, D., Helmer, C., Dartigues, J.F. 2009. Aluminum and silica in drinking water and the risk of Alzheimer's disease or cognitive decline: findings from 15-year follow-up of the PAQUID cohort. *Am. J. Epidemiol.* 169:489-96.

[65] Gillette-Guyonnet, S., Andrieu, F., Nourhashemi, S., de La Gueronniere, V., Grandjean, H. and Vellas, B. 2005. Cognitive impairment and composition of drinking water in women: findings of the EPIDOS Study. *Am. J. Clin. Nutr.* 81: 897-902.

[66] Zawilla, N.H., Taha, F.M., Kishk, N.A., Farahat, S.A., Farghaly, M., Hussein, M. 2014. Occupational exposure to aluminum and its amyloidogenic link with cognitive functions. *J. Inorg. Biochem.* 139: 57-64.

[67] Barage, S.H., Sonawane, K.D. 2015. Amyloid cascade hypothesis: Pathogenesis and therapeutic strategies in Alzheimer's disease. *Neuropeptides.* 52: 1-18.

[68] Wirths, O., Multhaup, G., Bayer, T.A. 2004. A modified ß-amyloid hypothesis: intraneuronal accumulation of the beta-amyloid peptide--the first step of a fatal cascade. *J. Neurochem.* 91: 513-20.

[69] Goate, A., Chartier-Harlin, M.C., Mullan, M., Brown, J., Crawford, F., Fidani, L. et al. 1991. Segregation of a missense mutation in the amyloid precursor protein gene with familial Alzheimer's disease. *Nature.* 349: 704-6.

[70] Sherrington, R., Rogaev, E.I., Liang, Y., Rogaeva, E.A., Levesque, G., Ikeda, M. et al. 1995. Cloning of a gene bearing missense mutations in early-onset familial Alzheimer's disease. *Nature.* 375: 754-60.

[71] Selkoe, D.J., Wolfe, M.S. 2007. Presenilin: running with scissors in the membrane. *Cell.* 131: 215-21.

[72] Yankner, B.A., Duffy, L.K., Kirschner, D.A. 1990. Neurotropic and neurotoxic effects of amyloid ß protein: reversal by tachykinin neuropeptides. *Nature.* 250: 279-82.
[73] Pike, C.J., Burdick, D., Walencewicz, A.J., Glabe C.G., and Cotman, C.W. 1993. Neurodegeneration induced by beta-amyloid peptides in vitro: the role of peptide assembly state. *J. Neurosci.* 13: 1676-87.
[74] Simmons, L.K., May, P.C., Tomaselli, K.J., Rydel, R.E., Fuson, K.S., Brigham, E.F. et al. 1994. Secondary structure of amyloid ß peptide correlates with neurotoxic activity in vitro. *Mol. Pharmacol.* 45: 373-9.
[75] Jarrett, J.T. and Lansbury, P.T. Jr. 1993. Seeding one-dimensional crystallization of amyloid: a pathogenic mechanism in Alzheimer's disease and scrapie? *Cell.* 73: 1055-8.
[76] Scheuner, D., Eckman, C., Jensen, M., Song, X., Citron, M., Suzuki, N. et al., 1996. Secreted amyloid ß-protein similar to that in the senile plaques of Alzheimer's disease is increased in vivo by the presenilin 1 and 2 and APP mutations linked to familial Alzheimer's disease. *Nat. Med.* 2: 864-70.
[77] Fukuyama, R., Mizuno, T., Mori, S., Nakajima, K., Fushiki S. and Yanagisawa, K. 2000. Age-dependent change in the levels of Aβ40 and Aβ42 in cerebrospinal fluid from control subjects, and a decrease in the ratio of Aβ42 to Aβ40 level in cerebrospinal fluid from Alzheimer's disease patients. *Eur. Neurol.* 43: 155-60.
[78] Kawahara, M. 2010. Role of calcium dyshomeostasis via amyloid channels in the pathogenesis of Alzheimer's disease. *Current Pharmaceutical Design.* 16: 2779-89.
[79] Dyrks, T., Dyrks, E., Masters, C.L., Beyreuther, K. 1993. Amyloidogenicity of rodent and human ß A4 sequences. *FEBS Lett.* 324: 231-6.
[80] Exley, C., Price, N.C., Kelly, S.M., and Birchall, J.D. 1993. An interaction of β-amyloid with aluminium *in vitro. FEBS Lett.* 324: 293-5.
[81] Mantyh, P.W., Ghilardi, J.R., Rogers, S., DeMaster, E., Allen, C.J., Stimson, E.R. et al. 1993. Aluminum, iron, and zinc ions promote aggregation of physiological concentrations of ß-amyloid peptide. *J. Neurochem.* 61: 1171-4.
[82] Kawahara, M., Muramoto, K., Kobayashi, K., Mori H., and Kuroda, Y. 1994. Aluminum promotes the aggregation of Alzheimer's β-amyloid protein *in vitro. Biochem. Biophys. Res. Commun.* 198: 531-5.
[83] Kawahara, M., Kato, M. and kuroda, Y. 2001. Effects of aluminum on the neurotoxicity of primary cultured neurons and on the aggregation of ß-amyloid protein. *Brain Res. Bull.* 55: 211-7.
[84] Bush, A.I., Pettingell, W.H., Multhaup, G., d Paradis, M., Vonsattel, J.P., Gusella, J.F. et al. 1994. Rapid induction of Alzheimer Aß amyloid formation by zinc, *Science.* 265: 1464-7.
[85] Atwood, C.S., Moir, R.D., Huang, X., Scarpa, R.C., Bacarra, N.M., Romano, D.M. et al. 1998. Dramatic aggregation of Alzheimer Aß by Cu(II) is induced by conditions representing physiological acidosis. *J. Biol. Chem.* 273: 12817-26.
[86] Bush, A.I. and Tanzi, R.E. 2008. Therapeutics for Alzheimer's disease based on the metal hypothesis. *Neurotherapeutics.* 5: 421-32.
[87] Kawahara, M., Negishi-Kato, M., Sadakane, Y. 2009. Calcium dyshomeostasis and neurotoxicity of Alzheimer's beta-amyloid protein. *Expert Rev. Neurother.* 9: 681-93.

[88] Hartley, D.M., Walsh, D.M., Ye, C.P., Diehl, T., Vasquez, S., Vassilev, P.M. et al. 1999. Protofibrillar intermediates of amyloid ß-protein induce acute electrophysiological changes and progressive neurotoxicity in cortical neurons. *J. Neurosci.* 19: 8876-84.

[89] Walsh, D.M., Tseng, B.P., Rydel, R.E., Podlisny, M.B., Selkoe, D.J. 2000. The oligomerization of amyloid ß-protein begins intracellularly in cells derived from human brain. *Biochemistry.* 39: 10831-9.

[90] Selkoe, D.J. 2008. Soluble oligomers of the amyloid beta-protein impair synaptic plasticity and behavior. *Behav. Brain Res.* 192: 106-13.

[91] Chen, W.T., Liao, Y.H., Yu, H.M., Cheng, I.H., Chen, Y.R. 2011. Distinct effects of Zn^{2+}, Cu^{2+}, Fe^{3+}, and Al^{3+} on amyloid-beta stability, oligomerization, and aggregation: amyloid-beta destabilization promotes annular protofibril formation. *J. Biol. Chem.* 286: 9646-56.

[92] Sharma, A.K., Pavlova, S.T., Kim, J., Kim, J., Mirica, L.M. 2013. The effect of Cu(2+) and Zn(2+) on the Aβ42 peptide aggregation and cellular toxicity. *Metallomics.* 5: 1529-36.

[93] Drago, D., Cavaliere, A., Mascetra, N., Ciavardelli, D., di Ilio, C. Zatta, P. et al. 2008. Aluminum modulates effects of beta amyloid(1-42) on neuronal calcium homeostasis and mitochondria functioning and is altered in a triple transgenic mouse model of Alzheimer's disease. *Rejuvenation Res.* 11: 861-71.

[94] Bolognin, S., Zatta, P., Lorenzetto, E., Valenti, M.T., Buffelli, M. 2013. β-Amyloid-aluminum complex alters cytoskeletal stability and increases ROS production in cortical neurons. *Neurochem. Int.* 62: 566-74.

[95] Muckenthaler, M.U., Galy, B., Hentze, M.W. 2008. Systemic iron homeostasis and the iron-responsive element/iron-regulatory protein (IRE/IRP) regulatory network. *Annu. Rev. Nutr.* 28: 197-213.

[96] Yamanaka, K., Minato, N., Iwai, K. 1999. Stabilization of iron regulatory protein 2, IRP2, by aluminum. *FEBS Lett.* 462: 216-20.

[97] Crichton, R.R., Wilmet, S., Legssyer, R., Ward, R.J. 2002. Molecular and cellular mechanisms of iron homeostasis and toxicity in mammalian cells. *J. Inorg. Biochem.* 91: 9-18.

[98] Oshiro, S., Kawahara, M., Kuroda, Y., Zhang, C., Cai, Y., Kitajima, S. et al. 2000. Glial cells contribute more to iron and aluminum accumulation but are more resistant to oxidative stress than neuronal cells. *Biochim. Biophys. Acta.* 1502: 405-14.

[99] Oshiro S., Kawahara M., Shirao M., Muramoto K., Kobayashi K., Ishige R. et al. 1998. Aluminum taken up by transferrin independent iron uptake affects the iron metabolism in rat cortical cells. *J. Biochem.* 123: 42-6.

[100] Oteiza, P.I. 1994. A mechanism for the stimulatory effect of aluminum on iron-induced lipid peroxidation. *Arch. Biochem. Biophys.* 308: 374-9.

[101] Kaneko, N., Sugioka, T., and Sakurai, H. 2007. Aluminum compounds enhance lipid peroxidation in liposomes: insight into cellular damage caused by oxidative stress. *J. Inorg. Biochem.* 101: 967-75.

[102] Everett, J., Céspedes, E., Shelford, L.R., Exley, C., Collingwood, J.F., Dobson, J. et al. 2014. Ferrous iron formation following the co-aggregation of ferric iron and the Alzheimer's disease peptide β-amyloid (1-42). *J. R. Soc. Interface.* 11:20140165. doi: 10.1098/rsif.2014.0165.

[103] Rogers, J.T., Bush, A.I., Cho, H.H., Smith, D.H., Thomson, A.M., Friedlich, A.L. et al. 2008. Iron and the translation of the amyloid precursor protein (APP) and ferritin mRNAs: riboregulation against neural oxidative damage in Alzheimer's disease. *Biochem. Soc. Trans.* 36: 1282-7.

[104] Wang, L., Hu, J., Zhao, Y., Lu, X., Zhang, Q., Niu, Q. 2014. Effects of aluminium on β-amyloid (1-42) and secretases (APP-cleaving enzymes) in rat brain. *Neurochem. Res.* 39: 1338-45.

[105] Liang, R.F., Li, W.Q., Wang, H., Wang, J.X., Niu, Q. 2013. Impact of sub-chronic aluminium-maltolate exposure on catabolism of amyloid precursor protein in rats. *Biomed. Environ. Sci.* 26:445-52.

[106] Walton, J.R., Wang, M.X. 2009. APP expression, distribution and accumulation are altered by aluminum in a rodent model for Alzheimer's disease. *J. Inorg. Biochem.* 103: 1548-54.

[107] Li, X.B., Zhang, Z.Y., Yin, L.H., Schluesener, H.J. 2012. The profile of β-amyloid precursor protein expression of rats induced by aluminum. *Environ. Toxicol. Pharmacol.* 33: 135-40.

[108] Lin, R, Chen, X., Li, W., Han, Y., Liu, P., Pi, R. 2008. Exposure to metal ions regulates mRNA levels of APP and BACE1 in PC12 cells: blockage by curcumin. *Neurosci. Lett.* 440: 344-7.

[109] Namekata, K., Imagawa, M., Terashi, A., Ohta, S., Oyama, F., Ihara, Y. 1997. Association of transferrin C2 allele with late-onset Alzheimer's disease. *Hum. Genet.* 101: 126-9.

[110] Imagawa, M., Naruse, S., Tsuji, S., Fujioka, A., Yamaguchi, H. 1992. Coenzyme Q10, iron, and vitamine B6 in genetically-confirmed Alzheimer's disease. *Lancet.* 340: 671.

[111] Frederickson, C.J., Suh, S.W., Silva, D., Frederickson, C.J., Thompson, R.B. 2000. Importance of zinc in the central nervous system: the zinc-containing neuron. *J. Nutr.* 130: 1471S-83S.

[112] Takeda, A., Nakamura, M., Fujii, H., Tamano, H. 2013. Synaptic Zn(2+) homeostasis and its significance. *Metallomics.* 5: 417-23.

[113] Weiss, J.H., Sensi, S.L., Koh, J.Y. 2000. Zn(2+): a novel ionic mediator of neural injury in brain disease. *Trends in Pharmacological Sciences.* 21: 395–401.

[114] Koyama, H., Konoha, K., Sadakane, Y., Ohkawara, S., and Kawahara, M. 2012. Zinc neurotoxicity and the pathogenesis of vascular-type dementia: Involvement of calcium dyshomeostasis and carnosine. *J. Clin. Toxicol.* S3-002 doi: 10.4172/2161-0495.S3-002.

[115] Kawahara, M., Mizuno, D., Koyama, H., Konoha, K., Ohkawara, S., Sadakane, Y. 2013. Disruption of zinc homeostasis and the pathogenesis of senile dementia. *Metallomics.* 6: 209-19.

[116] Singla, N., Dhawan, D.K. 2015. Zinc down regulates Apaf-1-dependent Bax/Bcl-2 mediated caspases activation during aluminium induced neurotoxicity. *Biometals.* 28: 61-73.

[117] Singla, N., Dhawan, D.K. 2014. Zinc modulates aluminium-induced oxidative stress and cellular injury in rat brain. *Metallomics.* 6: 1941-50. doi: 10.1039/c4mt00097h.

[118] Kawahara, M., Kato-Negishi, M., Hosoda, R., Kuroda, Y. 2002. Characterization of zinc-induced apoptosis of GT1-7 cells. *Biomed. Res. Trace Elements.* 13: 280-1.

[119] Green, K.N., LaFerla, F.M. 2008. Linking calcium to Aß and Alzheimer's disease. *Neuron*. 59: 190-4.

[120] Mattson, M.P., Chan, S.L. 2003. Neuronal and glial calcium signaling in Alzheimer's disease. *Cell Calcium*. 34: 385-97.

[121] Kawahara, M. 2004. Disruption of calcium homeostasis in the pathogenesis of Alzheimer's disease and other conformational diseases. *Curr. Alzheimer Res.* 1: 87-95.

[122] Büsselberg, D., Platt, B., Haas, H.L., Carpenter, D.O. 1993. Voltage gated calcium channel currents of rat dorsal root ganglion (DRG) cells are blocked by Al^{3+}. *Brain Res.* 622: 163-8.

[123] Rani, A., Neha, Sodhi, R.K., Kaur, A. 2015. Protective effect of a calcium channel blocker diltiazem on aluminum chloride-induced dementia in mice. *Naunyn Schmiedebergs Arch. Pharmacol.* 388: 1151-61.

[124] Achary, V.M., Parinandi, N.L., Panda, B.B. 2013. Calcium channel blockers protect against aluminium-induced DNA damage and block adaptive response to genotoxic stress in plant cells. *Mutat. Res.* 751: 130-8.

[125] Konoha, K., Imamura, L., Tsuda, M., Kawahara, M. 2003. Aluminum maltolate inhibits brain-derived neurotrophic factor induced intracellular free calcium elevations in cultured rat hippocampal neurons. *Biomed. Res. Trace Elements*. 14: 173-5.

[126] Carrell, R.W. and Lomas, D.A. 1997. Conformational disease. *Lancet*. 350: 134-8.

[127] Arispe, N., Rojas, E., Pollard, H.B. 1993. Alzheimer disease amyloid ß protein forms calcium channels in bilayer membranes: blockade by tromethamine and aluminum. *Proc. Natl. Acad. Sci. USA*. 90: 567-71.

[128] Lal, R., Lin, H., Quist, A.P. 2007. Amyloid ß ion channel: 3D structure and relevance to amyloid channel paradigm. *Biochim. Biophys. Acta*. 1768: 1966-75.

[129] Kawahara, M., Ohtsuka, I., Yokoyama, S., Kato-Negishi, M. and Sadakane, Y. 2011. Membrane incorporation, channel formation, and disruption of calcium homeostasis by Alzheimer's ß-amyloid protein. *Int. J. Alzheimer Dis.* 304583.

[130] Arispe, N., Pollard, H.B., and Rojas, E. 1996. Zn^{2+} interactions with Alzheimer's amyloid ß protein calcium channels. *Proc. Natl. Acad. Sci. USA*. 93: 1710-5.

[131] Kawahara, M., Arispe, N., Kuroda, Y., Rojas, E. 1997. Alzheimer's disease amyloid ß-protein forms Zn^{2+}-sensitive, cation-selective channels across excised membrane patches from hypothalamic neurons. *Biophys. J.* 73: 67-75.

[132] Kawahara, M., Arispe, N., Kuroda, Y., Rojas, E. 2000. Alzheimer's ß-amyloid, human islet amylin and prion protein fragment evoke intracellular free-calcium elevations by a common mechanism in a hypothalamic GnRH neuronal cell-line. *J. Biol. Chem.* 275: 14077-83.

[133] 133 Ciccotosto, G.D., James, S.A., Altissimo, M., Paterson, D., Vogt, S., Lai, B. et al. 2014. Quantitation and localization of intracellular redox active metals by X-ray fluorescence microscopy in cortical neurons derived from APP and APLP2 knockout tissue. *Metallomics*. 6: 1894-904.

[134] Wong, B.X., Tsatsanis, A., Lim, L.Q., Adlard, P.A., Bush, A.I., Duce, J.A. 2014. β-Amyloid precursor protein does not possess ferroxidase activity but does stabilize the cell surface ferrous iron exporter ferroportin. *PLoS One*. 9:e114174. doi: 10.1371/journal.pone.0114174. eCollection 2014.

[135] Crespo, Â.C., Silva, B., Marques, L., Marcelino, E., Maruta, C., Costa, S., et al. 2014. Genetic and biochemical markers in patients with Alzheimer's disease support a concerted systemic iron homeostasis dysregulation. *Neurobiol. Aging.* 35: 777-85.
[136] Singh, I., Sagare, A.P., Coma, M., Perlmutter, D., Gelein, R., Bell, R.D. et al. 2013. Low levels of copper disrupt brain amyloid-β homeostasis by altering its production and clearance. *Proc. Natl. Acad. Sci. USA.* 110: 14771-6.
[137] Prusiner, S.B. 1997. Prion diseases and the BSE crisis. *Science.* 278: 245–51.
[138] Mizuno, D., Koyama, H., Ohkawara, S., Sadakane, Y., Kawahara, M. 2014. Involvement of trace elements in the pathogenesis of prion diseases. *Curr. Pharam. Biotechnol.* 15:1049-57.
[139] Brown, D.R., Qin, K., Herms, J.W., Madlung, A., Manson, J., Strome, R. et al. 1997. The cellular prion protein binds copper in vivo. *Nature.* 390: 684-7.
[140] Jackson, G.S., Murray, I., Hosszu, L.L., Gibbs, N., Waltho, J.P., Clarke, A.R. et al. 2001. Location and properties of metal-binding sites on the human prion protein. *Proc. Natl. Acad. Sci. USA.* 98: 8531-5.
[141] Brown, D.R. 2009. Brain proteins that mind metals: a neurodegenerative perspective. *Dalton Trans.* 21: 4069-76.
[142] Schmitt-Ulms, G., Ehsani, S., Watts, J.C., Westaway, D., Wille, H. 2009. Evolutionary descent of prion genes from the ZIP family of metal ion transporters. *PLoS One.* 4:e7208.
[143] Fukada, T. and Kambe, T. 2011. Molecular and genetic features of zinc transporters in physiology and pathogenesis. *Metallomics.* 3:662-74.
[144] Watt, N.T., Griffiths, H.H., Hooper, N.M. 2013. Neuronal zinc regulation and the prion protein. *Prion.* 7: 203-8.
[145] Singh, A., Haldar, S., Horback, K., Tom, C., Zhou, L., Meyerson, H., et al. 2013. Prion protein regulates iron transport by functioning as a ferrireductase. *J. Alzheimers Dis.* 35: 541-52.
[146] Bobela, W., Aebischer, P., Schneider, B.L. 2015. Alpha-synuclein as a mediator in the interplay between aging and Parkinson's disease. *Biomolecules.* 5:.2675-700.
[147] Uversky, V.N., Li, J., Fink, A.L. 2001. Metal-triggered structural transformations, aggregation, and fibrillation of human alpha-synuclein. A possible molecular NK between Parkinson's disease and heavy metal exposure. *J. Biol. Chem.* 276: 44284-96.
[148] Davies, P., Moualla, D., Brown, D.R. 2011. α-synuclein is a cellular ferrireductase. *PLoS One.* 6: e15814.
[149] Cahill, C.M., Lahiri, D.K., Huang, X., Rogers, J.T. 2009. Amyloid precursor protein and alpha synuclein translation, implications for iron and inflammation in neurodegenerative diseases. *Biochim. Biophys. Acta.* 1790: 615-28.
[150] Schikorski, T., Stevens, C.F. 1997. Quantitative ultrastructural analysis of hippocampal excitatory synapses. *J. Neurosci.* 17: 5858-67.
[151] Ordóñez-Gutiérrez, L., Torres, J.M., Gavín, R., Antón, M., Arroba-Espinosa, A.I., Espinosa, J.C. et al. 2013. Cellular prion protein modulates β-amyloid deposition in aged APP/PS1 transgenic mice. *Neurobiol. Aging.* 34: 2793-804.
[152] Fluharty, B.R., Biasini, E., Stravalaci, M., Sclip, A., Diomede, L., Balducci, C. et al. 2013. An N-terminal fragment of the prion protein binds to amyloid-β oligomers and inhibits their neurotoxicity in vivo. *J. Biol. Chem.* 288: 7857-66.

[153] Han, H., Weinreb, P.H., Lansbury, P.T. Jr. 1995. The core Alzheimer's peptide NAC forms amyloid fibrils which seed and are seeded by beta-amyloid: is NAC a common trigger or target in neurodegenerative disease? *Chem. Biol.* 2: 163-9.
[154] Sindreu, C., Bayés, Á., Altafaj, X., Pérez-Clausell, J. 2014. Zinc transporter-1 concentrates at the postsynaptic density of hippocampal synapses. *Mol. Brain.* 7: 7-16.
[155] Sogawa, C.A., Asanuma, M., Sogawa, N., Miyazaki, I., Nakanishi, T., Furuta, H. et al. 2001. Localization, regulation, and function of metallothionein-III/growth inhibitory factor in the brain. *Acta Med. Okayama.* 55: 1-9.
[156] Bakardjiev, A., Bauer, K. 2000. Biosynthesis, release, and uptake of carnosine in primary cultures. *Biochemistry (Mosc.).* 65: 779-82.
[157] Kawahara, M., Konoha, K., Nagata, T., Sadakane, Y. 2007. Aluminum and human health: its intake, bioavailability and neurotoxicity. *Biomed. Res. Trace Elements.* 18: 111-20.
[158] Taylor, G.A., Ferrier, I.N., McLoughlin, I.J., Fairbairn, A.F., McKeith, I.G., Lett, D. et al. 1992. Gastrointestinal absorption of aluminium in Alzheimer's disease: response to aluminium citrate. *Age and Aging.* 21: 81-9.
[159] Lin, J.L., Yang, Y.J., Yang, S.S., Leu, M.L. 1997. Aluminum utensils contribute to aluminum accumulation in patients with renal disease. *Am. J. Kidney Dis.* 30: 653-8.
[160] Jouhanneau, P., Raisbeck, G.M., Yiou, F., Lacour, B., Banide, H., Drueke, T.B. 1997. Gastrointestinal absorption, tissue retention, and urinary excretion of dietary aluminum in rats determined by using ^{26}Al. *Clin. Chem.* 43: 1023-8.
[161] Kobayashi, K., Yumoto, S., Nagai, H., Hosoyama, Y., Imamura, M., Masuzawa, S. et al. 1990. ^{26}Al tracer experiment by accelerator mass spectrometry and its application to the studies for amyotrophic lateral sclerosis and Alzheimer's disease. I. *Proc. Japan. Acad. Ser. B* 66: 189-92.
[162] Bishop, N.J., Morley, R., Day, J.P., Lucas, A. 1997. Aluminum neurotoxicity in preterm infants receiving intravenous-feeding solutions. *N. Eng. J. Med.* 336: 1557-61.
[163] Federal Register 68: 32979-32981, 2003.
[164] McGraw, M. 1986. Aluminium content of milk formulate and intravenous used infants. *Lancet.* 1: 157.
[165] Yumoto, S., Nagai, H., Kobayashi, K., Tamate, A., Kakimi, S., Matsuzaki, H. 2003. ^{26}Al incorporation into the brain of suckling rats through maternal milk. *J. Inorg. Biochem.* 97:155-60.

In: Aluminum Neurotoxicity
Editor: Ankica Jelenković
ISBN: 978-1-63484-739-1

Chapter 7

Possible Neuroprotective Mechanism of Antioxidants against Aluminium Induced Cognitive Dysfunction

Anil Kumar*[*], *Arti Singh and Samrita Dogra
Pharmacology Division, University Institute of Pharmaceutical Sciences,
UGC Centre of Advanced Study, Panjab University, Chandigarh, India

Abstract

Aluminum (Al), the third most abundant nonessential element in Earth's crust, which has gained easy access to the life of humans. It is chemically bound to other elements, not typically bioavailable, so play no role in any known biochemistry of plants, animals or humans. Over the last decade, Al has become much more prevalent in the human environment. It is widely used in industries, processed foods, various medicinal compounds, as a flocculent in water treatment and as in the manufacturing of kitchen utensils and food and drink packaging, the production of dyes, baking powder and antacids. Because of such ubiquity, it is increasingly found in our bodies and we can say that today's age is termed as "The Aluminium Age". Aluminium accumulation in the brain is proposed to be associated with the pathophysiology of various neurodegenerative disorders, including Alzheimer's dementia, amylotrophic lateral sclerosis, Guam-Parkinson's dementia, etc. In this chapter, we will discuss about the mechanism of aluminium induced neurotoxicity along with the neuroprotective mechanism of various antioxidants.

Keywords: aluminium, neurodegenerative diseases, antioxidants, aluminium induced neurotoxicity

[*] Corresponding author: Dr. Anil Kumar, PhD, MBA; Professor of Pharmacology, Pharmacology division, University Institute of Pharmaceutical Sciences, UGC- Centre of Advanced Studies, Panjab University, Chandigarh, Email: kumaruips@yahoo.com.

INTRODUCTION

Aluminium (Al) is one of the most toxic metals studied so far and has been implicated in the etiology of many diseases (Kumar et al., 2015). It is also known to affect various organ systems along with major impact on the nervous system. In humans, Al plays a causative role in various disease conditions like dialysis encephalopathy, osteomalacia and microcytic anemia (Kumar and Gill, 2014). The toxic actions of Al on the nervous system include: disruption of synaptic activity, promotion of oxidative stress, misfolding of crucial proteins and increased permeability of the blood–brain barrier. Previous studies have shown that Al accumulation in the brain is being associated with the pathophysiology of neurodegenerative disorders, including Alzheimer's dementia, amylotrophic lateral sclerosis, Guam-Parkinson's dementia, etc. because of the reported prolonged half-life of Al (i.e., 150 days in rats) (Garruto et al., 1984; Perl et al., 1982; Candy et al., 1984; Kumar and Gill, 2014) (Figure 1). The overall impact of Al exposure showed neuropathological, neurobehavioral, neurophysical and neurochemical changes (Miu et al., 2003; Colomina et al., 2002; Kaur et al., 2006; Walton, 2012, 2013). Immunohistochemistry have conformed the accumulation of Al in the medial striatum, corpus callosum and cingulate bundle which further causing learning impairment (Platt et al., 2001). Because of the non-dividing nature of the neurons, the brain is considered to be the most vulnerable organ to the toxic exposure of Al owing. During Al induced neurotoxicity, there occurs oxidative stress in the brain, which further leads to increased level of free radicals and decreased level of antioxidants. The major property of Al is reported to have a strong prooxidant activity in spite of its nonredox status (Exley, 2004). Another study showed that, Al causes oxidative damage by binding to the metals (which are pro-oxidant in nature) like iron and copper (Zaky et al., 2013). It has also been reported that Al forms Al-superoxide anion complex, which is more potent oxidant than its superoxide anion and promotes further the formation of hydrogen peroxide and hydroxyl radicals and results in oxidative stress (Exley, 2004; Mahdi et al., 2006; Kumar and Gill, 2014).

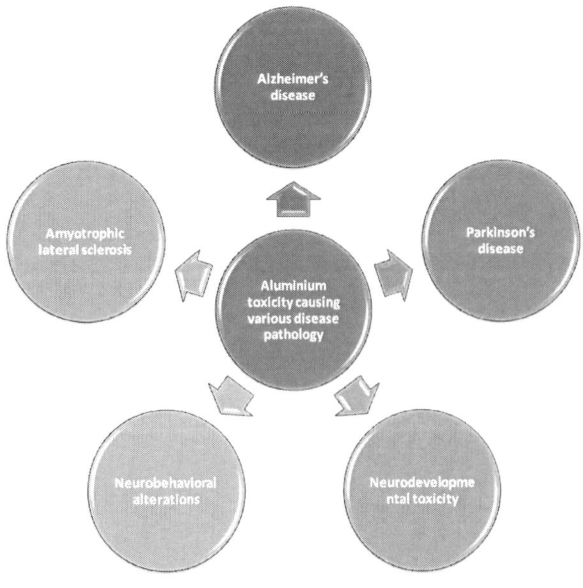

Figure 1. Flowchart showing aluminium toxicity induced various disease pathology.

GLOBAL PREVALENCE

As per WHO reports, humans get inevitably exposed to Al through food, cooking utensils, deodorants, antacids (Kaur and Gill, 2006; Kumar and Gill, 2014) apart from occupational exposure in the gun, automobile, aerospace and defense related factors (Sinczuk- Walczak et al., 2003; Sharma and Sharma, 2012), increasing the risk of neurodegenerative diseases such as Alzheimer's, Parkinsonism, etc. (Sharma and Sharma, 2012). It has been reported that the daily intake of Al is estimated to be approximately 10–20mg from different sources like cooking utensils, food additives, and medicines such as antacids, or deodorants (Edwardson et al., 1992).

MECHANISM OF ALUMINIUM INDUCED NEUROTOXICITY

There is no single unifying mechanism of Al induced neurotoxicity, but varieties of the cascade that has been identified over the last decade. It has been proposed that accumulation of Al accompanied by the release of cytochrome c from mitochondria, which eventually leads to increase in the production of free radicals causing oxidative stress by accelerating oxidative damage to biomolecules and an increase in the production of pro-inflammatory cytokines causing neuroinflammation (Kawahara, 2005). Further, it results in the increase in the gene expression of TNF-α, NF-κB and macrophage inflammatory protein-1 alpha (MIP-1α) (Johnson and Sharma, 2003).

Another mechanism of Al induced neurotoxicity reported was the depletion of the neuronal cells in the hippocampal and isocortical regions of the brain (Miu et al., 2006) along with the degeneration of cholinergic terminals in the cortical areas. It further accumulates in the cingulated bundle of the hippocampal regions of the brain and thereby induces learning deficits (Platt et al., 2001). Another study suggested that Al interferes with the glutamatergic neurotransmission (Provan et al., 1992) and impairs hippocampal long term potentiation by disrupting the glutamate-NO-cyclic guanosine monophosphate signaling pathway (Canales et al., 2001).

One of the most accepted hypotheses of Al induced neurotoxicity, is the model involving exacerbation of oxidative stress by enhancing superoxide and iron-driven oxidation (Gonc-Alves and Silva, 2007). Studies on animals showed that Al bio-accumulates to relatively high concentrations in the brain, readily complexes with basic brain biology and affects genes regulation. It further causes neuronal cell death and white matter erosion to reduce the connectivity of the brain (Miu et al., 2006). In 2004, Yang has reported that Al via neurotransmitters through cognitive and physical activity interact with the brain. They also reported that glial cell line-derived neurotrophic factor (GDNF) suppresses aluminum induced apoptosis of pyramidal neurons in rat brains, but brain derived neurotrophic factor (BDNF) may promote the apoptotic effect of aluminum (Yang et al., 2004).

Studies also suggest that Al acts as cholinotoxin by interacting with the cholinergic system (Kaizer et al., 2005), which further causes neuroinflammation by interacting with cholinergic projections functions and contributes to pathological processes that leads to learning and memory deficits (Exley, 2006). Further studies on Al showed that it gains access to the brain through transferrin mediated transport, which subsequently leads to neurotoxicity

(Yokel et al., 1999). Although, the exact mechanism behind aluminium induced neurotoxicity remains unclear, but the reports suggest increased oxidative stress and neuroinflammatory stress markers (Kumar et al., 2009) as the major hallmark responsible for disruption of intraneuronal metal homeostasis (Figure 2) (Julka and Gill, 1995) as well as axonal transport and long term potentiation (Wenting et al., 2014). Studies have shown that the brain consists of low levels of Al and its uptake in the brain is much slower as compared to other organs, but once gained access into the brain, Al distributes into the various regions. (Walker et al., 1994; Yokel, 2000; Jansson, 2001).

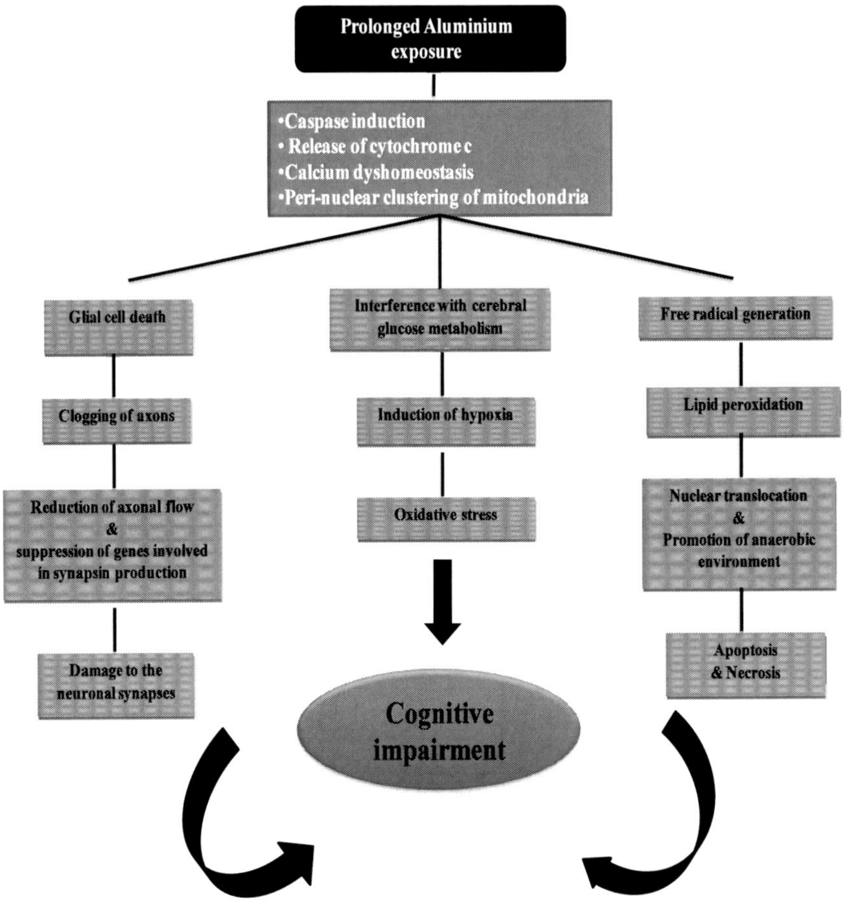

Figure 2. Possible mechanisms involved by which prolonged aluminium exposure result in cognitive dysfunction.

In-vivo studies in rodents showed that Al exposure causes disturbed cholinergic system and induction of neurobehavioral alterations in different brain regions (Yellamma et al., 2010; Kumar and Gill, 2009). Al accumulation further induces lipid peroxidation (LPO) in rat brain (Oteiza et al., 1993; Nehru and Anand, 2005) which is the result of ROS and reactive nitrogen species production and is an indirect measure of a metal's ability to promote oxidation. It has also been reported that Al being non-redox metal, may not alone produce an increase in oxidative damage (Bondy and Kirstein, 1996) but it can increase the Fe-induced production of ROS or membrane peroxidation (Xie et al., 1996; Ohyashiki et al., 1998; Campbell et al.,

1999; Xie and Yokel, 1996). Mechanism behind this was explained by Kim et al. in 2007, that Al stimulate uptake of nontransferrin bound iron and iron bound to transferrin in human glial cells in a dose and time-dependent manner (Kim et al., 2007).

In-vitro study on human brain cells in primary culture showed that neurotransmitters and their metabolites are very sensitive to Al exposure depicted that even a low dose of Al sulfate (100 nM) can induce pro-inflammatory and pro-apoptotic gene expression (Goncalves and Silva, 2007; Lukiw et al., 2005).

Another report correlates the development of neurological dysfunctions in patients of dialysis encephalopathy to high Al concentrations in the dialysate and to the use of phosphate binding gels containing Al (Jack et al., 1984). A questionnaire based study on the workers who were exposed to welding fumes containing Al and manganese reported that decreased motor function and chronic renal failure (Sjogren et al., 1996; Iregren et al., 2001; Shirabe et al., 2002). Al neurotoxicity from contaminated water at Camelford in the peoples showed considerable damage to cerebral function which may even lead to AD (Altmann et al., 1999; Neri and Hewitt, 1991; Gauthier et al., 2000; Flaten, 2001; Buchta et al. (2003). Neuropathological effects of Al toxicity include, nerve cell atrophy with stromal spongiosis, proliferation of astrocytes and microglia in the cerebral cortex, basal ganglia and thalamus. Another study on the neurotoxic effects of Al among foundry workers showed early neurotoxic symptoms of AD and Al has also been observed in hippocampal neurons from humans with AD (Polizzi et al. 2002; Walton, 2006, 2009). Recently, a study showed significantly higher levels of Al and silver in the hair of the newborns with neural tube defects imposed that these metals may be participating as factors in the development of neural tube defects (Ramirez-Altamirano et al. (2012).

Treatment Strategies against Al Induced Neurotoxicity

Amelioration of Aluminium Neurotoxicity by Chelation Therapy

With the advancing age risk of the occurrence of acute and chronic human intoxications has been considerably increased from the past couple of years with a wide range of metals. Studies have confirmed the effectiveness of the chelation therapy based on the toxicokinetics and chemical considerations of metal and chelating agents. Basically chelating agents works primarily on the mobilization of toxic metal from the body, resulting in reduction of metal toxicity, by forming a stable complex with a toxic metal may shield the metal ion from biological targets, thereby reducing the toxicity, or it may expose the metal to the biological environment and prevent the metal from being scavenged by biological protective mechanisms and thereby increase the toxicity of the metal (Andersen, 1999). A study on desferrioxamine or deferoxamine (DFO), an iron chelator, shows that it prevents iron catalyzed free radical reactions (Tilbrook and Hider, 1998), has been used in dialysis-related Al toxicity syndromes because of the high stability of the DFO–Al complex (Day and Ackrill, 1993), reduced tissue Al concentrations after subacute loading in rabbits, reverse Al induced LPO, suggesting its probable therapeutic role against Al neurotoxicity (Julka and Gill, 1996), partially reverse Al induced neurofibrillary degeneration in rabbits (Savory et al., 1994).

Chelators of trivalent cations following oral Al loading in rats and showed that oral 1,2-dimethyl-3-hydroxypyrid-4-one was the most effective in enhancing urinary Al excretion Gomez et al. (1999) and these types of chelators were also more effective in reducing brain Al in Al-loaded rats (Florence et al., 1995).

Amelioration of Aluminium Neurotoxicity by Antioxidants

Antioxidants are substances, used to inhibit or delay the oxidation of a substrate. In Al induced neurotoxicity induced oxidative stress and mitochondrial dysfunction with rebalancing the impaired prooxidant/antioxidant ratio has been studied through supplementation of many antioxidants (Kumar and Singh, 2015). It is evident from the literature that there are various agents that provides significant protective effects as antioxidants, including some chelators, plant extracts and antioxidant nutrients such as vitamins, selenium, zinc, curcumin, melatonin and so forth. The most important source of antioxidants is provided by nutrition. Antioxidant molecules are thought to play a crucial role in counteracting free radical induced damage to macromolecules. *Vitamin E,* lipid soluble chain-breaking antioxidant, known to protect biological membranes and lipoproteins from oxidative stress (Packer, 1991). The major biological function includes its direct influence on cellular responses to oxidative stress through modulation of signal transduction pathways (Azzi et al., 1992). A study on male rats, shows that concomitant dietary vitamin E supplementation to Al lactate (10 mg/kg b.wt. i.p. 4 weeks) decreased brain tissue Al levels in different regions and increased CAT and GSH levels (Abubakar et al. (2004). Another major antioxidant, *vitamin C*, that scavenges the aqueous ROS by very rapid electron transfer that inhibits LPO (Halliwell and Gutteridge, 1989) and also known to be effective in the protection of Al-induced toxicity in different tissues of rabbits (Yousef, 2004).

Another potent scavenger of ROS, *melatonin,* that prevents the reduction of membrane fluidity caused by LPO (Garcia et al., 1997), scavenges free radicals by neutralizing peroxyl or free hydroxyl radicals (Tan et al., 1993; Pieri et al., 1994), reduces lipid and protein oxidation in a concentration-dependent manner in isolated synaptosomes treated with Al chloride (Albendea et al. (2007). A study on experimental rats shows that pretreatment with melatonin protects against chronic Al chloride induced TBARS accumulation in liver, kidney and brain indicating that orally administered melatonin prior to Al chloride is probably responsible for the prevention of oxidative damage and Al-toxicity (Fyiad (2007). A study on rabbit brain treated with Al shows that exogenous melatonin also exerts potential protection from Al lactate induced oxidative stress in cortex and cerebellum (abed-Elghaffar et al., 2005). Another study on female transgenic (Tg2576) and wild-type mice fed with Al lactate showed that melatonin exerted antioxidant action by increasing the mRNA levels of SOD1, CAT, and GR antioxidant enzymes (Garcia et al., 2010).

Selenium is a cofactor of GPx, which enhances the availability of GSH, which is one of the most abundant intrinsic antioxidant involved in preventing LPO and resultant cell damage. Another study, including a combination of *vitamin E and selenium* as joint antioxidants in rats administered with 34 mg/kg b.wt. Al chloride for 30 days showed significant decrease in oxidative stress (El-Demerdash, 2004), restore the adverse effects on the intracellular oxidative stress processes in the liver (Viezeliene et al., 2013). *Zinc* an essential micronutrient plays an important role in the DNA replication, transcription, and protein synthesis, influencing cell division and differentiation (Frederickson, 1989). Also reported as an essential component of the oxidant defense system and attenuated DNA damage (Sato and Bremner, 1993; Singla and Dhawan (2013). Zinc is also known to decreased terminal TUNEL-positive cells, the number of apoptotic brain cells, p53 and nuclear factor kB levels after Al treatment hence possess inflammatory properties Singla and Dhawan (2013). *Lithium,* has been reported to modulate the apoptosis-regulatory proteins present in the hippocampus of rabbit brain (Ghribi et al., 2002), also exhibit neuroprotective role following Al exposure Bhalla and Dhawan (2009).

N-acetyl cysteine, a synthetic precursor of reduced GSH is a thiol-containing compound, which stimulates the intracellular GSH synthesis, enhances GST activity and acts as a scavenger of free radicals. The administration of N-acetyl cysteine has been shown to restore the increased LPO and altered antioxidant defense system in the brain when rats were administrated with Al chloride (100 mg/kg) for 6 weeks through oral gavages (Prakash and Kumar, 2009). *S-allyl-cysteine*, a bioactive and a bioavailable component of garlic, scavenges free radicals and reducing LPO and enzymatic alteration inflicted by sub chronic Al nitrate administration in rats (Shrivastava, 2011).

Curcumin has been shown to possess potent antioxidant, antitumor promoting and anti-inflammatory properties *in-vitro* and *in-vivo*. It is known to attenuate oxidative stress and protective effects in tissues and can potentially reduce Al induced adverse effects in brain. Curcumin supplementation is known to be effective in normalizing the activities of mitochondrial complex I, II and IV as well as reduced GSH content following chronic Al administration in rats (Sood et al., 2011). Similarly, chronic curcumin administration along with Al improves memory retention, attenuates oxidative damage and reduces Al accumulation in rats as demonstrated by Kumar et al. (2009a). Sharma et al. (2009) also presented the evidence that curcumin exerts a protective effect against Al-induced elevation of ageing-related changes by modulating the extent of oxidative stress as evident by up regulation in the activities of antioxidant enzymes.

Role of Plant Extracts and Bioflavonoids

Literature review has shown that plant extracts and flavonoids were used as effective strategies as antioxidants in Al induced neurotoxicity. It has been reported that the traditional Indian medicinal system provides the varieties of antioxidants like *Bacopa monniera*, a nerve tonic, which possesses antioxidative, adaptogenic and memory enhancing effects. A study on rat brain treated Al showed that chronic treatment with *Bacopa monniera* extract and donepezil restored the altered antioxidant enzyme activities. It also attenuates damage to Purkinje's neurons and altered granular cell layer along with the increased accumulation of intranueronal lipofuscin granules and necrotic alteration in the CA1 region of the hippocampus which occurs due to Al (Tripathi et al., 2011). *Bacopa monniera* extract also prevented the decrease in SOD activity as well as the increase in oxidative damage to lipids and proteins in cerebral cortex and hippocampus of rat brain treated with Al treatment (Jyoti and Sharma, 2006). Another study on pomegranate peel showed that its antioxidant properties decreased Al accumulation and stimulated anti-apoptotic proteins against Al exposure in rat brain (Abdel Moneim, 2012). A study on Fisetin showed that it can attenuate increased LPO and reduction in GSH in chronic Al chloride treated mice (Prakash et al. 2013b). Fenugreek seeds ameliorated enhanced LPO and altered lipid profile in various organs of rats including brain (Belaid-Nouira et al., 2012).

A flavanone glycoside, naringin (for 6 weeks at dose of 100 mg/kg through oral gavages), has been reported to restore the mitochondrial functions in the rat brain treated with Al chloride (Prakash et al., 2013a). Another natural flavonoid, quercetin, has been reported to ameliorate chronic Al exposure induced neurodegenerative changes, reduced oxidative stress, decreased ROS production, increased MnSOD activity and glutathione levels with decreased LPO and protein oxidation as observed by Al accumulation (Sharma et al., 2013b). Another falvanoid, resveratrol, exerts anti-inflammatory action by inhibiting different pro-inflammatory cytokines and key signalling molecules in rat brain treated with Al (Zaky et al., 2013). Chronic *Centella asiatica* treatment also showed improved memory performance, oxidative defense, decreased Al accumulation and reversed mitochondrial enzyme activity in rats treated with Al (Prakash and Kumar (2013). The saffron extract alone or in combination with honey syrup has shown the antioxidant activity that could reverse the Al-induced changes in the levels of MDA and GSH in Al treated mice (Linardaki et al. 2013; Shari et al., 2011). *Celastrus paniculatus,* seed extract, also reported known to improve rodent memory functions against Al chloride induced neural insult, oxidative stress (Chakrabarty et al., 2012) and improvement in antioxidant enzyme activities and decrease in the level of LPO or DNA oxidation (Sumathi et al., 2013).

N-nitro-L-arginine methyl ester, a nonspecific nitric oxide synthase inhibitor, has been reported to exert anti-oxidant actions in Al chloride treated animals by decreasing nitrite concentrations, superoxide anion production, MDA levels and reduced GSH contents in the forebrain cortex of Al treated rats (Stevanovic et al., 2009). *Caffeic acid*, a specific inhibitor of 5-lipoxygenase, may have potential value in the treatment of cerebral damage induced in mice by intracerebroventricular Al (Yang et al., 2008). *In-vitro* and *in-vivo* studies of necrostatin-1 on neural cell of mice brain showed neuroprotective role induced by Al exposure (Qinli et al., 2013).

Miscellaneous Approaches against Al Induced Neurotoxicity

In-vivo study on rats showed that chronic administration of *carvedilol* along with Al exposure significantly improved the memory performance tasks in the Morris water maze test, attenuated oxidative stress by reducing LPO, restored reduced GSH and SOD, CAT and GST activity in Al-treated rat brain (Kumar et al., 2011). Administration of a traditional medicine, *Manasamitra vatakam*, used for the treatment of psychosomatic diseases, anxiety neurosis and stress also improves memory performance and antioxidant activity against Al chloride induced neurotoxicity in rats (Thirunavukkarasu et al., 2012). A non-steroidal anti-inflammatory drug *meloxicam*, well known inhibitor of cyclooxygenase, showed protection from cerebral neuronal damage, decreased LPO as well as normalization of cyclooxygenase-2 mRNA in Al-overload mice (Jun-Qing et al., 2006; Nivsarkar et al., 2006). An anti-ageing drug, *centrophenoxine,* used to treat senile dementia in AD patients, has also been reported to restore altered redox status and improving the thiol status in rat brain regions induced Al neurotoxicity (Nehru and Bhalla, 2006a). It also attenuates the altered LPO, reduced GSH content or antioxidant enzyme activities in rat brain regions following post treatment to Al (Nehru and Bhalla, 2006b; Nehru et al., 2007). Another agent, *Ginkgo biloba,* has been reported to reversed the neurotoxic effects of sub chronic Al exposure in rat brain by attenuating oxidized GSH, MDA as well as reduced GSH (Abd-Elhady et al., 2013).

Drug Combination Strategies

Combination strategies are always much better than the single drug administration and various studies have reported that co-administration of dietary nutrients like vitamins or essential metal with a chelating agent provide more beneficial effects than chelator or antioxidants alone (Flora and Tandon, 1990; Tandon et al., 1994; Lauwerys et al., 1983). A study on rat showed that, N (2-hydroxyethyl) ethylenediamine triacetic acid (HEDTA) and selenium co-administration for 7 consecutive days to the brain after 60 days exposed to Al improved reduced GSH, GR, GPx, CAT and SOD activities and increase in LPO and GST levels which was significant as compared to HEDTA alone (Shrivastava, 2012). Another combination of HEDTA with propolis decreased LPO and increased the activity of ATPase, SOD, CAT and GSH contents in hepatic, renal and nervous tissues which were altered by a single dose of Al nitrate administration when compared with the monotherapy (Bhadauria, 2012). Previous study on HEDTA and citric acid combination along with Al nitrate treatment showed better responses to alterations in GST levels, brain TBRS levels and GPx activity as compared to their effect alone. (Flora et al., 2003). Another study showed that Tiron chemically known as 4,5- dihydroxybenzene-1,3-disulfonic acid disodium salt when combined with GSH produced more beneficial effects against decrease in reduced GSH, GR, GPx, CAT, SOD activities and elevation in TBARS and GST when compared with individual treatment induced by the Al chloride administration in rat brain (Sharma et al., 2007). So, combination strategies are far much better than single drug therapy.

CONCLUSION

In this chapter, we have discussed the no evidences that show Al as a neurotoxic metal which can exacerbate oxidative stress neuroinflammation. To date, various mechanisms of Al neurotoxicity have been proposed, but the best is yet to come. It has also been discussed that along with oxidative stress, mitochondrial dysfunction is also considered to be the major hallmark of Al neurotoxicity. In Al induced neurotoxicity, functional impairment of mitochondria causes ROS generation, oxidative damage to mitochondrial proteins and DNA and finally causes mitochondrial dysfunction. This mechanism indicated the involvement of Al in the pathogenesis of neurodegenerative diseases such as AD, PD and ALS. The research on Al needs further investigation, apart from its involvement in the generation of oxidative stress and mitochondrial dysfunction, in terms of duration of exposure of the Al. Various antioxidants, chelating agents and plant derived products either alone or in combination ameliorates Al induced oxidative stress and mitochondrial dysfunction.

REFERENCES

Abdel, M.A.E., (2012). Evaluating the potential role of pomegranate peel in aluminum induced oxidative stress and histopathological alterations in brain of female rats. *Biol. Trace. Elem. Res.* 150, 328–336.

Abd-Elghaffar, S.K.H., El-Sokkary, G.H., Sharkawy, A.A., (2005). Aluminum-induced neurotoxicity and oxidative damage in rabbits: protective effect of melatonin. *Neuro. Endocrinol. Lett.* 26, 609–616.

Abd-Elhady, R.M., Elsheikh, A.M., Khalifa, A.E., (2013). Anti-amnestic properties of Ginkgo biloba extract on impaired memory function induced by aluminum in rats. *Int. J. Dev. Neurosci.* 31, 598–607.

Abubakar, M.G., Taylor, A., Ferns, G.A., (2004). Regional accumulation of aluminium in the rat brain is affected by dietary vitamin E. *J. Trace. Elem. Med. Biol.* 18, 53–59.

Albendea, C.D., Gomez-Trullen, E.M., Fuentes-Broto, L., Miana-Mena, F.J., Millan-Plano, S., Reyes-Gonzales, M.C. et al., (2007). Melatonin reduces lipid and protein oxidative damage in synaptosomes due to aluminium. *Trace. Elem. Med. Biol.* 21, 261–268.

Altmann, P., Cunningham, J., Dhanesha, U., Ballard, M., Thompson, J., Marsh, F., (1999). Disturbance of cerebral function in people exposed to drinking water contaminated with aluminium sulphate: retrospective study of the Camelford water incident. *BMJ.* 319, 807–811.

Andersen, O., (1999). Principles and recent developments in chelation treatment of metal intoxication. *Chem. Rev.* 99, 2683–2710.

Azzi, A., Boscoboinik, D., Hensey, C., (1992). The protein kinase C family. *Eur. J. Biochem.* 208, 547–557.

Becaria, A., Lahiri, D.K., Bondy, S.C., Chen, D., Hamadeh, A., Li, H et al., (2006). Aluminum and copper in drinking water enhance inflammatory or oxidative events specifically in the brain. *J. Neuroimmunol.* 176, 16–23.

Belaid-Nouira, Y., Bakhta, H., Bouaziz, M., Flehi-Slim, I., Haouas, Z., Ben, C.H., (2012). Study of lipid profile and parieto-temporal lipid peroxidation in AlCl3 mediated neurotoxicity. Modulatory effect of fenugreek seeds. *Lipids Health Dis.* 11, 16.

Bhadauria, M., (2012). Combined treatment of HEDTA and propolis prevents aluminum induced toxicity in rats. *Food Chem. Toxicol.* 50, 2487–2495.

Bhalla, P., Dhawan, D.K., (2009). Protective role of lithium in ameliorating the aluminium-induced oxidative stress and histological changes in rat brain. *Cell. Mol. Neurobiol.* 29, 513–521.

Bondy, S.C., Kirstein, S., (1996). The promotion of iron-induced generation of reactive oxygen species in nerve tissue by aluminium. *Mol. Chem. Neuropathol.* 27, 185–194.

Buchta, M., Kiesswetter, E., Otto, A., Schaller, K.H., Seeber, A., Hilla, W et al., (2003). Longitudinal study examining the neurotoxicity of occupational exposure to aluminium containing welding fumes. *Int. Arch. Occup. Environ. Health.* 76, 539–558.

Campbell, A., Hamai, D., Bondy, S.C., (2001). Differential toxicity of aluminium salts in human cell lines of neural origin: implications for neurodegeneration. *Neurotoxicology.* 22, 63–71..

Canales, J.J., Corbalàn, R., Montoliu, C., Llansola, M., Monfort, P., Erceg, S., Hernandez-Viadel, M., Felipo, V., (2002). Aluminium impairs the glutamate-nitric oxide-cGMP pathway in cultured neurons and in rat brain in vivo: molecular mechanisms and implications for neuropathology. *J. Inorgan. Biochem.* 87, 63-69.

Candy, J.M., Oakley, A.E., Atack, J., Perry, R.H., Perry, E.K., Edwardson, J. A., (1984). New observations on the nature of senile plaque cores. *Regulation Transmitter Funct.* 17, 301-304.

Chakrabarty, M., Bhat, P., Kumari, S., D'Souza, A., Bairy, K.L., Chaturvedi, A et al., (2012). Corticohippocampal salvage in chronic aluminium induced neurodegeneration by Celastrus paniculatus seed oil: neurobehavioural, biochemical, histological study. *J. Pharmacol. Pharmacother.* 3, 161–171.

Colomina, M.T., Roig, J.L., Sa´nchez, D.J., Domingo, J.L., (2002). Influence of age on aluminum-induced neurobehavioral effects and morphological changes in rat brain. *Neurotoxicology.* 23, 775–781.

Day, J.P., Ackrill, P., (1993). The chemistry of desferrioxamine chelation for aluminum overload in renal dialysis patients. *Ther. Drug. Monit.* 6, 598–601.

Exley, C., Esiri, M.M., (2006). Severe cerebral congophilic angiopathy coincident with increased brain aluminium in a resident of Camelford, Cornwall, UK. J. Neurol. Neurosurg. *Psychiatry.* 77(7), 877-879.

El-Demerdash, F.M., (2004). Antioxidant effect of vitamin E and selenium on lipid peroxidation, enzyme activities and biochemical parameters in rats exposed to aluminium. *J. Trace. Elem. Med. Biol.* 18, 113–121.

Flaten, T.P., (2001). Aluminium as a risk factor in Alzheimer's disease, with emphasis on drinking water. *Brain. Res. Bull.* 55, 187–196.

Flora, S.J., Mehta, A., Satsangi, K., Kannan, G.M., Gupta, M., (2003). Aluminum-induced oxidative stress in rat brain: response to combined administration of citric acid and HEDTA. *Comp. Biochem. Physiol. C. Toxicol. Pharmacol.* 134:319–328.

Flora, S.J.S., Tandon, S.K., (1990). Beneficial effects of zinc supplementation during chelation treatment of lead intoxication in rats. *Toxicology.* 64, 129–139.

Florence, A.L., Gauthier, A., Ward, R.J., Crichton, R.R., (1995). Influence of hydroxypyridones and desferrioxamine on the mobilization of aluminium from tissues of aluminium loaded rats. *Neurodegeneration.* 4, 449–455.

Frederickson, C.J., (1989). Neurobiology of zinc and zinc containing neurons. *Int Rev Neurobiol.* 31, 145–238.

Fyiad, A.A., (2007). Aluminium toxicity and oxidative damage reduction by melatonin in rats. *J. Appl. Sci. Res.* 3, 1210–1217.

Garcia, J.J., Reiter, R.J., Guerrero, J.M., Escames, G., Yu, B.P., Oh, C.S., et al., (1997). Melatonin prevents changes in microsomal membrane fluidity during induced lipid peroxidation. *FEBS Lett.* 408, 297–300.

Gauthier, E., Fortier, I., Courchesne, F., Pepin, P., Mortimer, J., Gauvreau, D., (2000). Aluminum forms in drinking water and risk of Alzheimer's disease. *Environ. Res.* 84, 234–246.

Ghribi, O., Herman, M.M., Spaulding, N.K., Savory, J., (2002). Lithium inhibits aluminum-induced apoptosis in rabbit hippocampus, by preventing cytochrome c translocation, Bcl-2 decrease, Bax elevation and caspase-3 activation. *J. Neurochem.* 82, 137–145.

Garruto, R.M., Fukatsu, R., Yanagihara, R., Gajdusek, D.C., Hook, G., Fiori, C.E.. (1984). Imaging of calcium and aluminium neurofibrillary tangle neurons in Parkinsonism-dementia of Guam. *Proc. Natl. Acad. Sci.* USA. 81 1875-1879.

Gomez, M., Esparza, J.L., Domingo, J.L., Singh, P.K., Jones, M.M., (1999). Chelation therapy in aluminum loaded rats: influence of age. *Toxicology. 137, 161–168.*

Goncalves, P.P., Silva, V.S., (2007). Does neurotransmission impairment accompany aluminium neurotoxicity? *J. Inorg. Biochem.* 101, 1291–1338.

Gorsky, J.E., Dietz, A.A., Spencer, H., Osis, D., (1979). Metabolic balance of aluminium studied in six men. *Clin. Chem.* 25, 1739–1743.

Halliwell, B., Gutteridge, J.M.C., (1989). *Free radicals in biology and medicine*. Oxford: Clarendon Press, 96–98.

Iregren, A., Sjogren, B., Gustafsson, K., Hagman, M., Nyle′n, L., Frech, W., et al., (2001). Effects on the nervous system in different groups of workers exposed to aluminium. *Occup. Environ. Med.* 58, 453–460.

Jansson, E.T., (2001). Aluminum exposure and Alzheimer's disease. *J. Alzheimers. Dis.* 3, 541–549.

Julka, D., Gill, K.D., (1996). Effect of aluminum on regional brain antioxidant defense status in Wistar rats. *Res. Exp. Med.* 196, 187–194.

Jun-Qing, Y., Bei-Zhong, L., Bai-Cheng, H., Qi-Qin, Z., (2006). Protective effects of meloxicam on aluminum overload-induced cerebral damage in mice. *Eur. J. Pharmacol.* 547, 52–58.

Jyoti, A., Sethi, P., Sharma, D., (2007). Bacopa monniera prevents from aluminium neurotoxicity in the cerebral cortex of rat brain. *J. Ethnopharmacol.* 111, 56–62.

Johnson, V.J., Sharma, R.P., (2003). Aluminium disrupts the pro-inflammatory cytokine/neurotrophin balance in primary brain rotation-mediated aggregate cultures: possible role in neurodegeneration. *Neurotoxicology.* 24, 261-268.

Julka, D., Gill, K.D., (1996). Altered calcium homeostasis: a possible mechanism of aluminium-induced neurotoxicity. *Biochim. Biophys. Acta.* 1315(1), 47-54.

Kaizer, R.R., Corrêa, M.C., Spanevello, R.M., (2005). Acetylcholinesterase activation and enhanced lipid peroxidation after long-term exposure to low levels of aluminium on different mouse brain regions. *J. Inorg. Biochem.* 99, 1865-1870.

Kawahara, M., (2005). Effects of aluminium on the nervous system and its possible link with neurodegenerative diseases. *J. Alz. Dis.* 8, 171-182.

Kim, M.K., Choi, G.J., Lee, H.S., (2003). Fungicidal property of Curcuma longa L. rhizome-derived curcumin against phytopathogenic fungi in a greenhouse. *J. Agric. Food. Chem. 51*(6), 1578-1581.

Kumar, A., Prakash, A., Dogra, S., (2011). Neuroprotective effect of carvedilol against aluminium induced toxicity: possible behavioral and biochemical alterations in rats. *Pharmacol. Rep.* 63, 915–923.

Kumar, V., Bal, A., Gill, K.D., (2009b). Aluminium-induced oxidative DNA damage recognition and cell-cycle disruption in different regions of rat brain. *Toxicology.* 264, 137–144.

Kumar, V., Bal, A., Gill, K.D., (2009c). Susceptibility of mitochondrial superoxide dismutase to aluminium induced oxidative damage. *Toxicology.* 255, 117–123.

Kumar, V., Gill, K.D., (2009). Aluminium neurotoxicity: neurobehavioural and oxidative aspects. *Arch. Toxicol.* 83, 965–978.

Kumar, V., Gill, K.D., (2014). Oxidative stress and mitochondrial dysfunction in aluminium neurotoxicity and its amelioration: A review. *NeuroToxicol.* 41, 154–166.

Kumar, A., Singh, A., Ekavali., (2015). A review on Alzheimer's disease pathophysiology and its management: an update. *Pharmacol. Rep.* 67, 195–203.

Kumar, A., Singh, A., (2015). A review on mitochondrial restorative mechanism of antioxidants in Alzheimer's disease and other neurological conditions. *Front. Pharmacol.* 6(206), 1-13.

Lauwerys, R., Roels, H., Buchet, J.P., Bernard, A.A., Verhoeven, L., Konings, J., (1983). The influence of orally-administered vitamin C or zinc on the absorption of and the biological response to lead. *J. Occup. Med.* 25, 668–678.

Linardaki, Z.I., Orkoula, M.G., Kokkosis, A.G., Lamari, F.N., Margarity, M., (2013). Investigation of the neuroprotective action of saffron (Crocus sativus L.) in aluminum-exposed adult mice through behavioral and neurobiochemical assessment. *Food Chem. Toxicol.* 52, 163–170.

Lukiw, W.J., Percy, M.E., Kruck, T.P., (2005). Nanomolar aluminum induces pro-inflammatory and pro-apoptotic gene expression in human brain cells in primary culture. *J. Inorg. Biochem.* 99, 1895–1898.

Mahdi, A.A., Tripathi, S., Neerja, J., Mahdi, H., (2006). Aluminium mediated oxidative stress: possible relationship to cognitive impairment of Alzheimers type. *Ann. Neurosci.* 13.

Miu, A.C., Benga, O., (2006). Aluminium and Alzheimer's disease: a new look. *J. Alz. Dis.* 10, 179-201.

Nehru, B., Bhalla, P., Garg, A., (2007). Further evidence of centrophenoxine mediated protection in aluminium exposed rats by biochemical and light microscopy analysis. *Food Chem. Toxicol.* 45, 2499-2505.

Nehru, B., Anand, P., (2005). Oxidative damage following chronic aluminium exposure in adult and pup rat brains. *J. Trace. Elem. Med. Biol.* 19, 203–208.

Neri, L.C., Hewitt, D., (1991). Aluminium, Alzheimer's disease, and drinking water. *Lancet.* 338, 390.

Nivsarkar, M., Banerjee, A., Shah, D., Trivedi, J., Patel, M., Cherian, B., et al., (2006). Reduction in aluminum induced oxidative stress by meloxicam in rat brain. *Iran. Biomed. J.* 10, 151–155.

Ohyashiki, T., Suzuki, S., Satoh, E., Uemori, Y., (1998). A marked stimulation of Fe2+-initiated lipid peroxidation in phospholipid liposomes by a lipophilic aluminum complex, aluminum acetylacetonate. *Biochim. Biophys. Acta.* 1389, 141–149.

Oteiza, P.I., Keen, C.L., Han, B., Golub, M.S., (1993). Aluminum accumulation and neurotoxicity in Swiss- Webster mice after long-term dietary exposure to aluminum and citrate. Metabolism. 42, 1296–1300.

Packer, L., (1991). Protective role of vitamin E in biological systems. *Am. J. Clin. Nutr.* 53, 1050–1055.

Pieri, C., Marra, M., Moroni, F., Recchioni, R., Marcheselli, F., (1994). Melatonin, a peroxide radical scavenger more effective than vitamin E. *Life Sci.*55, 271–276.

Platt, B., Fiddler, G., Riedel, G., Henderson, Z., (2001). Aluminium toxicity in the rat brain: histochemical and immunocytochemical evidence. *Brain. Res. Bull.* 55, 257–267.

Polizzi, S., Pira, E., Ferrara, M., Bugiani, M., Papaleo, A., Albera, R., et al. (2002). Neurotoxic effects of aluminium among foundry workers and Alzheimer's disease. *Neurotoxicology.* 23, 761–774.

Perry, C.C., Keeling-Tucker, T., (1998). Aspects of the bioinorganic chemistry of silicon in conjunction with the biometals calcium, iron and aluminium. *J. Inorg. Biochem.* 69, 181–191.

Prakash, A., Kumar, A., (2013). Mitoprotective effect of Centella asiatica against aluminuminduced neurotoxicity in rats: possible relevance to its anti-oxidant and antiapoptosis mechanism. *Neurol. Sci.* 34, 1403–1409.

Prakash, A., Shur, B., Kumar, A., (2013a). Naringin protects memory impairment and mitochondrial oxidative damage against aluminum-induced neurotoxicity in rats. *Int. J. Neurosci.* 123, 636–645.

Prakash, D., Gopinath, K., Sudhandiran, G., (2013b). Fisetin enhances behavioral performances and attenuates reactive gliosis and inflammation during aluminum chloride-induced neurotoxicity. *Neuromol. Med.* 15, 192–198.

Qinli, Z., Meiqing, L., Xia, J., Li, X., Weili, G., Xiuliang, J., et al. (2013). Necrostatin-1 inhibits the degeneration of neural cells induced by aluminum exposure. *Restor. Neurol. Neurosci.* 31:543–555.

Ramirez-Altamirano, M.J., Fenton-Navarro, P., Sivet-Chinas, E., Harp-Iturribarria, F.M., Martinez-Cruz, R., Cruz, P.H., et al. (2012). The relationship of aluminium and silver to neural tube defects; a case control. *Iran. J. Pediatr.* 22, 369–374.

Sato, M., Bremner, I., (1993). Oxygen free radicals and metallothionein. *Free. Radic. Biol. Med.* 14, 325–337.

Savory, J., Huang, Y., Herman, M.M., Reyes, M.R., Wills, M.R., (1995). Tau immunoreactivity associated with aluminum maltolate-induced neurofibrillary degeneration in rabbits. *Brain Res.* 669, 325–329.

Sharma, D., Sethi, P., Hussain, E., Singh, R., (2009). Curcumin counteracts the aluminium-induced ageing-related alterations in oxidative stress, Na+, K+ ATPase and protein kinase C in adult and old rat brain regions. *Biogerontology.* 10, 489–502.

Sharma, D.R., Sunkaria, A., Wani, W.Y., Sharma, R.K., Kandimalla, R.J., Bal, A., et al. (2013). Aluminium induced oxidative stress results in decreased mitochondrial biogenesis via modulation of PGC-1a expression. *Toxicol. Appl. Pharmacol.*

Sharma, D.R., Wani, W.Y., Sunkaria, A., Kandimalla, R.J., Verma, D., Cameotra, S.S., (2013b). Quercetin protects against chronic aluminum-induced oxidative stress and ensuing biochemical, cholinergic, and neurobehavioral impairments in rats. *Neurotoxicol. Res.* 23, 336–357.

Shirabe, T., Irie, K., Uchida, M., (2002). Autopsy case of aluminium encephalopathy. *Neuropathology.* 22, 206–210.

Shrivastava, S., (2012). Combined effect of HEDTA and selenium against aluminum induced oxidative stress in rat brain. *J. Trace. Elem. Med. Biol.* 26, 210–214.

Shrivastava, S., (2011). S-allyl-cysteines reduce amelioration of aluminum induced toxicity in rats. *Am. J. Biochem. Biotechnol.*7, 74–83.

Singla, N., Dhawan, D.K., (2013). Zinc, a neuroprotective agent against aluminum-induced oxidative DNA injury. *Mol. Neurobiol.* 48, 1–12.

Sjogren, B., Iregren, A., Frech, W., Hagman, M., Johansson, L., Tesarz, M., et al. (1996). Effects on the nervous system among welders exposed to aluminium and manganese. *Occup. Environ. Med.* 53, 32–40.

Sood, P.K., Nahar, U., Nehru, B., (2011). Curcumin attenuates aluminum-induced oxidative stress and mitochondrial dysfunction in rat brain. *Neurotox. Res.* 20, 351–361.

Stevanovic, I.D., Jovanovic, M.D., Colic, M., Ninkovic, M., Jelenkovic, A., Mihajlovic, R., (2013) Cytochrome C oxidase activity and nitric oxide synthase in the rat brain following aluminium intracerebral application. *Folia. Neuropathol.* 51, 140–146.

Sumathi, T., Shobana, C.S., Mahalakshmi, V.R., Sureka, R.C., Subathra, M., Vishali, A., et al. (2013). Oxidative stress in brains of male rats intoxicated with aluminium and neuromodulating effect of Celastrus paniculatus alcoholic seed extract. *Asian. J. Pharm. Clin. Res.* 6, 80–90.

Tan, D.X., Chen, L.D., Poeggler, B., Manchester, L.C., Reiter, R.J., (1993). Melatonin: a potent endogenous hydroxyl radical scavenger. *Endocr. J.* 1, 57–60.

Tandon, S.K., Singh, S., Jain, V.K., (1994). Efficacy of combined chelation in lead intoxication. *Chem. Res. Toxicol.* 7, 585–589.

Thirunavukkarasu, S.V., Venkataraman, S., Raja, S., Upadhyay, L., (2012). Neuroprotective effect of Manasamitra vatakam against aluminium induced cognitive impairment and oxidative damage in the cortex and hippocampus of rat brain. *Drug. Chem. Toxicol.* 35, 104–115.

Tilbrook, G.S., Hider, R.C., (1998). Iron chelators for clinical use. *Met. Ions. Biol. Syst.* 35, 691– 730.

Tripathi, S., Mahdi, A.A., Hasan, M., Mitra, K., Mahdi, F., (2011). Protective potential of Bacopa monniera (Brahmi) extract on aluminum induced cerebellar toxicity and associated neuromuscular status in aged rats. *Cell. Mol. Biol.* 57, 3–15.

Viezeliene, D., Beekhof, P., Gremmer, E., Rodovicius, H., Sadauskiene, I., Jansen, E., et al. (2013). Selective induction of IL-6 by aluminum-induced oxidative stress can be prevented by selenium. *J. Trace. Elem. Med. Biol.* 27, 226–229.

Walker, V.R., Sutton, R.A., Meirav, O., Sossi, V., Johnson. R., Klein, J., et al. (1994). Tissue disposition of aluminum in rats measured by accelerator mass spectrometry. *Clin. Invest. Med.* 17, 420–425.

Walton, J.R., (2012). Aluminum disruption of calcium homeostasis and signal transduction resembles change that occurs in aging and Alzheimer's disease. *J. Alzheimers. Dis.* 29, 255–273.

Walton, J.R., (2013). Aluminum involvement in the progression of Alzheimer's disease. *J. Alzheimers. Dis.* 35, 7–43.

Xie, C.X., Yokel, R.A., (1996). Aluminum facilitation of iron-mediated lipid peroxidation is dependent on substrate, pH, and aluminium and iron concentrations. *Arch. Biochem. Biophys.* 327, 222–226.

Xie, C.X., Mattson, M.P., Lovell, M.A., Yokel, R.A., (1996). Intraneuronal aluminum potentiates iron induced oxidative stress in cultured rat hippocampal neurons. *Brain. Res.* 743, 271–277.

Yang, J.Q., Zhou, Q.X., Liu, B.Z., He, B.C., (2008). Protection of mouse brain from aluminum-induced damage by caffeic acid. *CNS Neurosci. Ther.* 14, 10–16.

Yang, S.J., Lee, J.E., Lee, K.H., Huh, J.W., Choi, S.Y., Cho, S.W., (2004). Opposed regulation of aluminuminduced apoptosis by glial cell line-derived neurotrophic factor and brain-derived neurotrophic factor in rat brains. *Brain. Res. Mol. Brain. Res.* 127, 146–149.

Yellamma, K., Saraswathamma, S., Kumari, B.N., (2010). Cholinergic system under aluminium toxicity in rat brain. *Toxicol. Int.* 17, 106–112.

Yokel, R.A., O'Callaghan, J.P., (1998). An aluminum-induced increase in GFAP is attenuated by some chelators. *Neurotoxicol. Teratol.* 20, 55–60.

Yokel, R.A., (2000). The toxicology of aluminum in the brain: a review. *Neurotoxicology,* 21, 813–828.

Yousef, M.I., (2004). Aluminium-induced changes in hemato-biochemical parameters, lipid peroxidation and enzyme activities of male rabbits: protective role of ascorbic acid. *Toxicology,* 199, 47–57.

Zaky, A., Mohammad, B., Moftah, M., Kandeel, K.M., Bassiouny, A.R., (2013). Apurinic/apyrimidinic endonuclease 1 is a key modulator of aluminum-induced neuroinflammation. *BMC Neurosci.* 14, 26.

About the Editor

Ankica Jelenković, MD, PhD
Clinical Pharmacologist
Principal Research Fellow
Institute for Biological Research "Siniša Stanković",
University of Belgrade, Belgrade, Republic of Serbia
Tel:/Fax +381 11 3420606
E-mail: jelaka@yahoo.com

Dr. Ankica Jelenković earned both an MD and PhD and has a specialization in clinical pharmacology. She has worked as a teaching assistant at the Medical Faculty, University of Niš, Department of Pharmacology and Toxicology. After that, and until today, she works at the Institute for Biological Research "Siniša Stanković", Belgrade, University of Belgrade, Republic of Serbia. Dr. Jelenković has written relevant scientific titles. Her basic scientific interest is directed to animal models of neurological diseases in humans. In addition to pathogenic mechanisms, research emphasis is placed over the treatment and prevention of these diseases by using drugs and other medications. In the field of clinical pharmacology, adverse drug reactions are her greatest field of interest.

INDEX

#

A

D

E

F

G

H

I

K

L

M

N

O

P

Q

R

S

T

V

W

X

Z

α

ß

γ